Algrove Publishing Limited
1090 Morrison Drive
Ottawa, Ontario
Canada K2H 1C2

Canadian Cataloguing in Publication Data

Clark, George H. (George Harold), 1872-1935?
Farm weeds of Canada

2nd ed.
(Classic reprint series)
Reprint of the ed. published Ottawa: Dept. of Agriculture, 1906.
ISBN 0-921335-81-4

1. Weeds--Canada. 2. Weeds--Control--Canada. I. Criddle, Norman. II. Fletcher, James, 1852-1908. III. Title. IV. Series: Classic reprint series (Ottawa, Ont.)

SB613.C2C53 2000 632′.5′0971 C00-900162-X

Printed in Canada
#20101

Publisher's Note

Although originally called *Farm Weeds of Canada,* this book actually covers the entire temperate belt of North America since weeds have respect for growing conditions only, not for borders. Some new weeds have been introduced to North America since this book was first published but it still covers virtually all those of interest to farmers and gardeners alike.

Leonard G. Lee, Publisher
Ottawa
March, 2000

DOMINION OF CANADA
Department of Agriculture
BRANCH OF THE SEED COMMISSIONER

Farm Weeds of Canada

BY

GEORGE H. CLARK, B.S.A.

AND

JAMES FLETCHER, LL.D., F.R.S.C., F.L.S.

WITH ILLUSTRATIONS BY

NORMAN CRIDDLE

PUBLISHED BY DIRECTION OF
THE HON. SYDNEY A. FISHER, MINISTER OF AGRICULTURE
OTTAWA, 1906

To the Honourable Sydney A. Fisher,

Minister of Agriculture,
Ottawa, Canada:

Sir,—I have the honour to submit herewith a special bulletin on the farm weeds of Canada, prepared in accordance with your instructions. The text of this bulletin is by Dr. James Fletcher, Entomologist and Botanist to the Dominion Experimental Farms, and the illustrations were made in water-colour from actual specimens, under his supervision by Mr. Norman Criddle, of Aweme, Man., while employed by the Seed Branch.

Incalculable losses are annually sustained by farmers on account of the prevalence of noxious weeds. The cost of the labour needed to control and eradicate them has become a serious problem in farm management.

There are many means by which weeds become disseminated. Weed seeds are carried by the waters of rivers and creeks. Transportation companies, especially the railway companies, are responsible for the introduction of many new weeds. The wind and animals of various kinds do much to spread weed seeds in a locality. The results of investigation work conducted by this Branch clearly show that the commerce in agricultural seeds is an exceedingly fruitful medium for the distribution of weeds from field to field and from province to province.

In the Seed Laboratory all kinds of seeds are tested free of charge to farmers. The reports issued thereon give the names of weed seeds found in the samples sent for test, but the mere names of weeds frequently have no meaning to persons who are not expert in the nomenclature of plants. In consequence most new weeds become introduced and well established before their noxious character is appreciated by farmers.

The reference collections of correctly named specimens of weed seeds that are prepared in the Seed Laboratory for distribution to seed merchants and agricultural institutions, are used in the identification of impurities found in commercial seeds. The illustrations of weeds and the information contained in this bulletin will make these reference collections of weed seeds more intelligible and serviceable.

Since the work of testing seeds for farmers and seed merchants was undertaken by this Branch, many applications have been received for a bulletin containing illustrations of the weeds named in the Seed Control Act and other weeds, the seeds of which were reported to be present in the samples sent for test. It was largely with a view to meet this need on the part of farmers that I suggested to you the preparation of this illustrated bulletin on farm weeds. I recommend that it be printed and distributed, free of charge, on personal application, for use as a reference book in the libraries of farm homes and rural schools.

It is a pleasant duty to record the hearty cooperation that I have at all times received from Dr. James Fletcher in compiling reference collections of weed seeds and other work of this Branch, but particularly am I grateful to him for his kindly co-operation and the deep interest he has taken in the preparation of this bulletin.

I have the honour to be, Sir,

Your obedient servant,

GEO. H. CLARK,

Seed Commissioner.

FARM WEEDS OF CANADA

By James Fletcher, Ottawa.

INTRODUCTORY.

The annual losses due to the occurrence of pernicious weeds upon farm lands, although acknowledged in a general way, are far greater than is actually realized. Where, however, a proper course of treatment based upon an accurate knowledge of the nature of each weed is adopted, these losses can be lessened very appreciably.

The present time seems to be particularly opportune for urging the need of systematic and united effort on the part of all engaged in the cultivation of the soil, in striving by every means in their power to fight against the increasing prevalence of many weeds of the farm. The recent enactment of the Seed Control Act and the very considerable losses to the farmers of the Western Provinces owing to the presence of such a large percentage of foul seeds in the bountiful crop of 1905, have awakened a keen interest, which, it is hoped, will induce a closer study of those general principles which affect the question of the introduction, spread and development of all weeds of the farm, as well as of the methods by which even the worst of these may be eradicated. These methods are all founded on a knowledge of the individual nature of each kind of weed, and there is no weed known which cannot be controlled and cleared from farm lands that are cultivated as they ought to be, with a suitable rotation of crops and with the ordinary implements now in use by Canadian farmers.

The subject of farm weeds and their eradication is now one of burning interest to all cultivators of the soil in every part of the Dominion. This interest is shown by frequent enquiries for the correct names and nature of any strange plants found growing among crops, and for advice as to the best means of controlling them. During the past ten years several official bulletins on weeds have been issued and widely distributed. In all of these publications the same names are given for the different weeds. It is, therefore, clearly important that those for whose benefit the bulletins have been prepared, should know the plants treated of by the names there used, so that they may be able to make the fullest use of the information supplied.

In the fight against noxious weeds, the first thing of importance is to know a weed when seen and to call it by its true name, not necessarily its botanical name, but the name by which it is generally known and written about in agricultural publications. Local names, unfortunately, are very often wrong. There are, for instance, at least half a dozen plants of quite different habits, which are locally known under the name of "Russian Thistle"; "Ragwort" is a name applied to several plants; "Black Mustard", again, is used for two or three troublesome plants of the Mustard family, whereas the true Black Mustard is seldom seen in Canada, and has nowhere appeared as a farm pest. "Chicory", "Milkweed" and "Bindweed" are names applied to many different plants. It cannot be made too widely known that anyone wishing to learn the names or nature of plants found on his land can send specimens post free to the Botanist of the Experimental Farm, at Ottawa; and full information about the plant will be forwarded to the sender gladly and free of all charge, with as little delay as possible. It is always better to send specimens when making enquiries, because so many weeds are locally known by wrong names. Therefore, if information is asked about a certain plant under a wrong name, it is very probable that the treatment suggested may not be suitable.

Farmers give very little critical attention to the different weeds growing among their crops. Some think that, because these plants are in a measure unfamiliar, the exact recognition of all of them is a matter beyond their power. This, however, is by no means the case, and, as the different kinds vary greatly in their powers of robbing the farmer, it is certainly advisable that more attention should be given to weed pests than has been done in the past. Although there are several hundreds of different kinds of plants growing wild in almost every locality and many of these may sometimes appear among cultivated crops, there are only a few which a farmer need trouble about—not many more than there are different kinds of crops grown; and every cultivator of the soil knows the difference between wheat, barley, oats, rye, peas, turnips, beets, etc. It is no more difficult, if the importance of the subject is recognized, to learn the names, nature and appearance at different stages of growth, and also the seeds of Stinkweed, Hare's-ear Mustard, False-flax, Canada Thistle, Field Sowthistle, Sweet Grass, Quack, etc., than to recognize the familiar plants which have been grown for many years as crops.

What is a Weed?

There are many definitions of the word Weed, but perhaps from a farmer's standpoint the best one is: "any injurious, troublesome, or unsightly plant that is at the same time useless or comparatively so." As a general statement, it may be said that our most troublesome and aggressive weeds of the farm have been introduced into Canada from other countries; but it is also true that under special circumstances some of our native wild plants may increase and become "noxious weeds."

Losses due to Weeds.

That the losses due to the presence of weeds upon cultivated land are enormous, is generally understood; but it may not be amiss here to draw special attention to some of the ways in which these enemies injure the tiller of the soil.

1. Weeds do great harm by robbing the soil of the plant food intended for the crop and also of its moisture, thus increasing the effects of drought by taking up water from the soil and wasting it by evaporation.

2. Weeds crowd out and take the place of more useful plants, being hardier, and, as a rule, more prolific.

3. Weeds are a source of great loss. From the time farmers begin to prepare their land for a crop, these enemies increase the cost of every operation—in ploughing, harrowing, seeding, cultivating, cutting, binding, carrying and threshing, as well as in cleaning, freighting and marketing the produce. Direct losses are the larger consumption of binder twine necessary when weedy grain crops are harvested, and the extra wear and tear on machinery due to coarse-growing weeds.

4. The eradication of the worst weeds is costly in labour, time and machinery and frequently compels a farmer to change what would be the best rotation of crops for the land, or even to grow crops which are not the most advantageous.

5. Many weeds are conspicuous and all are unsightly on farm lands. They thus, in a varying degree according to their several natures, depreciate the value of land, a point which may be of great importance, should the owner wish to sell.

6. Some weeds are injurious to stock, being poisonous as the Water Hemlock, or injurious to their products as burs in wool, or Wild Garlic and Stinkweed, which taint milk. The horny or barbed seeds of some grasses cause irritation or painful wounds by penetrating the flesh and particularly the mouth parts, as Porcupine Grass and Skunk-tail Grass in the North-west.

7. Weeds attract injurious insects or harbour fungous diseases, which may spread to cultivated plants. It is well known that weedy stubbles or summer-fallows are a breeding ground for cutworms and that the rust of small grains may pass the winter on several kinds of grasses.

In view of the above, it must be acknowledged that in all parts of Canada weeds are a source of constant and very considerable loss to farmers. Indeed, so much is this the case that the great prevalence of some varieties in certain districts of the Dominion must be viewed with the gravest alarm, for they have taken such possession of the land as to seriously affect profitable farming. As examples of such aggressive enemies, mention may be made of the Wild Mustard, Quack or Couch Grass and Canada Thistle in parts of almost every province; Ox-eye Daisy in the Maritime Provinces; Field Sowthistle in the Maritime Provinces, Quebec, Ontario and the Red River valley in Manitoba; and Stinkweed or Penny-cress, Ball Mustard and Hare's-ear Mustard in all the Prairie Provinces.

The increase of weeds in a district has frequently been due to the fact that farmers have neglected them from not being aware of their noxious nature and power to spread.

"Many of our farmers have only a limited knowledge of weeds, and in many cases do not recognize those that are dangerous on their first appearance. Hence we have 'One year's seeding, seven years weeding.' There are some weeds so noxious that, if farmers knew their real character and recognized the plants on their first appearance, they would postpone all other business until they were destroyed. * * * Self-interest should be a sufficient incentive to farmers to destroy weeds if it is clearly shown that it will pay them to do so."—H. MacKellar.

A point of considerable importance with regard to noxious weeds is the adoption, as much as possible, of some one English or common name. The names used in this bulletin have been selected with much care as being those which are most applicable and most widely known. When more names than one are given, the first is preferable. The scientific names, of which only one for each plant is recognized as authoritative by botanists all over the world, are here given, so that the certain identity of each plant mentioned may be known. Few farmers, of course, are acquainted with these scientific terms, even in the case of our commonest weeds, but it would be well if they were; for certainly much confusion exists in different localities in the application even of the English popular names, the same plant being frequently called by one name in one place and by quite a different name elsewhere, or quite as frequently a single name is applied to a number of distinct plants in different places or by different people in the same place.

How Weeds Spread.

In the present age of extensive and easy communication with all parts of the country, and indeed with the whole world, there are frequent opportunities for seeds of weeds being introduced into previously uninfested districts.

There are many ways in which weeds are spread:—

1. By natural agencies. The wind carries the seeds long distances not only in summer but with drifting soil and over the surface of the snow in winter. Streams distribute them far and wide along their courses. They are also distributed by seed-eating birds and herbivorous animals, through the stomachs of which the seeds have passed undigested, or by being attached to some part of their bodies by special contrivances, with which nature has provided some seeds for this very purpose, such as hooked and barbed hairs, spines, gummy excretions, etc.

2. By human agency. The seeds of weeds are frequently introduced as "foul seed" mixed with other seeds, particularly in cheap, improperly cleaned seed; they are also brought on to previously clean farms with manure from towns or are imported in hay used for packing or as fodder. In addition to this, they are often distributed over farms by waggons, harrows, seeders, threshing machines or other agricultural implements. As an offset against the great benefits we derive from railways, it has been found that many bad weeds have been introduced into new localities through their agency, the seeds being either shaken from cars or cleaned out of them at stopping places.

A fact, however, which should not be forgotten, is that the railway companies do not grow these weed seeds themselves. They are merely the carriers for farmers and would far rather handle grain perfectly free from weed seeds of all kinds than such as contains a possible source of injury to a new district, in the prosperity and progress of which they are interested. It is most important to keep a close watch on all railway banks and station yards so as to detect and destroy any new weeds which may appear, before they spread to the surrounding country.

Classification of Weeds.

Weeds, like all other plants, may be simply classified according to the length of time they live, under the three following heads:—Annuals, or one-year plants; Biennials, or two-year plants; and Perennials, or many-year plants. In eradicating weeds, it is of the greatest importance to consider under which of these heads they come, because in most instances the treatment is simple and will be upon the general principles of preventing annuals and biennials from seeding, and perennials from forming new leaves, roots and underground stems.

Annuals.—Are those plants which complete their whole growth in a year. As a rule they have small fibrous roots and produce a large quantity of seed. Examples of this class are found in Wild Mustard, Stinkweed, Lamb's-quarters, Wild Buckwheat, Purslane, Ragweed, Wild Oats. There are also some annuals called "Winter Annuals," which are not only true annuals when the seeds germinate in spring, but are also biennial in habit, that is, their seeds ripened in the summer fall to the ground, germinate and produce a certain growth before winter sets in and then complete their development the next spring. Of these may be mentioned Shepherd's-purse, Peppergrass, Stinkweed, mentioned above, Wormseed Mustard, Ball Mustard, Hare's-ear Mustard, Canada Flea-bane, and the Blue Bur.

Biennials.—Are those plants which require two seasons to complete their growth, the first being spent in collecting and storing up a supply of nourishment, which is used the second season in producing flowers and seeds. Examples of these are Burdock, False-tansy, Common Evening-primrose and Viper's Bugloss or Blue-weed.

Perennials—Are those plants which continue growing for many years. Perennial weeds are propagated in several ways, but all produce seeds as well. They have two distinct modes of growth, those which root deeply, and those of which the the root system is near the surface. The most troublesome are those which extend long underground stems deep beneath the surface of the ground, as Canada Thistle, Perennial or Field Sowthistle, Field Bindweed, Bladder Campion, White-stemmed Evening-primrose, Showy Lettuce, and some wild Sunflowers. Representatives of the second class or shallow-rooted perennials are: Pasture Sage, Yarrow and Couch Grass. Some perennials extend but slowly from the root by means of short stems or offsets, but produce a large quantity of seed. Of these, Ox-eye Daisy, Dandelion, Goldenrod and Yarrow are examples.

Extermination of Weeds.

In adopting a method of extermination, the nature of the plant to be eradicated and its habits of growth must, first of all, be taken into consideration.

Annuals.—Any method by which the germination of the seed in the soil is hastened and the young plants afterwards destroyed before they produce fresh seed, must in time clean land however badly infested it may be with annual weeds. The seeds of some annuals have great vitality, and will continue appearing for several years as fresh seeds are brought up to the surface by cultivation. Wild Mustard seeds have been known to germinate after lying deep in the ground for twenty years, and the same vitality has been claimed for Wild Oats, with, however, less satisfactory evidence.

Biennials must be either ploughed up or cut down before they flower. Mowing at short intervals in the second year, so as to prevent the formation of new seed, will clear the land of this class of plants; but a single mowing will only induce them to send out later branches, which, if not cut, will mature many seeds. Where ploughing is impracticable, this class of plants should be cut off below the crown of the root. For this purpose, a spud or a large chisel at the end of a long handle (to obviate the necessity of stooping) is as convenient a tool as can be used.

Perennials are by far the most troublesome of all weeds and require very thorough treatment, in some instances the cultivation of special crops, to ensure their eradication. Imperfect treatment such as a single ploughing, frequently does more harm than good, by breaking up the underground stems and stimulating growth.

It will be found in examining several perennial plants that they may be divided into two classes, one of which has its root system close to the surface of the soil, while the other roots very deeply. For the shallow-rooted perennials, infested land should be either trenched deeply or ploughed so lightly that the roots are exposed to the sun to dry up. For deep-rooted perennials, on the other hand, ploughing should be as deep as conveniently possible without going to extra expense. The depth of ploughing must be decided by the nature of the land. In light or gravelly soils shallow ploughing may be preferable, as too deep ploughing would interfere with the mechanical texture of the soil, which is so important in the movement of soil moisture.

Some method of cultivation must be adopted, such as the use of a broad-shared cultivator, to cut off at short intervals all freshly formed shoots an inch or two beneath the surface of the soil, so as to prevent the plants from forming leaves and thus storing up nourishment to sustain future growth.

Plants take in most of their food through their leaves. Perennial plants, which live for many years, have special reservoirs where some of this food after elaboration is stored up for future use in such receptacles as bulbs, tubers and fleshy rootstocks. The first growth in spring, particularly flowering stems, are produced mainly by drawing on this special store of nourishment. Plants are therefore in their weakest condition, at that time of the year when they have exhausted to the greatest extent their supply of reserve food, and have not had time to replenish it. The stage of growth, then, when ploughing down of perennial plants will be of most effect, is just when their flowering stems have made their full growth, but before the seeds, which would be a source of danger, have had time to mature. Some experience is necessary to know what is the best time to work certain soils or to deal with special weeds, as well as to recognize weeds in all their stages. Some weeds, the Russian Thistle and Stinkweed for instance, have a very different appearance when young and when mature. No general rule can be given, as the necessary treatment will vary in different districts on different soils and under different climatic conditions. What may be the proper treatment in one place, may fail in another. Perennial plants, if allowed to develop flower stems and then ploughed down (or first mowed and then ploughed under), will, by the production of the flower stems, have so far reduced the nourishment stored up in the rootstocks that they are much weakened and can afterwards be easily dealt with by two or three cultivations before winter sets in. Late fall ploughing has also been found extremely useful in Manitoba in cleaning land infested by some of the worst perennial weeds, such as Canada Thistle and Field Sowthistle. On the other hand, it has been found in the West that all the weeds and other plants decay readily if prairie lands or meadows are broken in May or early in June. Land so treated can therefore be cleaned far more easily than if the operation of breaking is delayed until July. This is due to the climate and to the succulent nature of all parts of the plant at that season.

General Principles.

1. *There is no weed known which cannot be eradicated by constant attention, if only the nature of its growth be understood.*

2. Never allow weeds to ripen seed.

3. Cultivate frequently, particularly early in the season, so as to destroy seedlings while small and easily killed.

4. Many weed seeds can be induced to germinate in autumn by cultivating stubbles immediately after harvest. Many of the seedlings would be killed by winter or could be easily disposed of by ploughing or cultivation in spring.

5. All weeds bearing mature seeds should be burnt, and under no circumstances should they be ploughed under.

6. All weeds can be destroyed by the use of the ordinary implements of the farm, the plough, the cultivator, the spud and the hoe.

7. Be constantly on the alert to prevent new weeds from becoming established on farms. Notwithstanding all efforts to the contrary, weeds will certainly be introduced from time to time even on the farms of the most careful.

Summer-Fallowing.

Summer-fallowing as an agricultural practice, although not now followed to the same extent as formerly in the older provinces, owing to the more general adoption of mixed farming with a short rotation, in which hoed crops are most useful in cleaning the land of weeds, is essentially necessary in those parts of Manitoba, Saskatchewan and Alberta, where the conservation of moisture in the soil is of great importance, the farms are large, labour is scarce and the time for cultivating the land in autumn and spring is limited.

The method of summer-fallowing recommended is: to plough deeply (so as to get a suitable seed bed) early in summer as soon after seeding as possible; harrow the same day the ploughing is done, so as to hold in the largest amount of moisture, and then prevent a growth of weeds on this land by three or at most four cultivations before winter sets in. This may be done with any kind of cultivator or a disk harrow.

It is recommended for most parts of the West to summer-fallow land once in three years. Plough and cultivate the first year. Grow grain the second year and, except in the driest districts, take one crop more off stubble. Then summer-fallow again. One ploughing of summer-fallow gives the best results, because in wet years a second deep ploughing will tend to produce too much growth and delay the ripening of the grain. Crops grown on stubble do not yield quite so well as on summer-fallow, but the grain ripens earlier and in windy sections there is less danger of the soil drifting. A second crop should never be sown on stubble. There has been a tendency in some parts of the West to put off the ploughing of land to be summer-fallowed, as long as possible, so as to reduce the subsequent labour of harrowing and cultivating. The danger, however, of ploughing down seeds of several kinds of winter annuals, which in the dry climate of Manitoba and the North-western Provinces are sufficiently developed to ripen after being ploughed down, is so great that, as recommended by Mr. Angus Mackay, the work should be begun directly after seeding is finished.

Short Rotation of Crops.

As a means of keeping farms free of weeds there are few methods of working land which give such good results as a systematic short rotation of crops with regular seeding down to grass or clover at short intervals.

In the Prairie Provinces mixed farming has not yet been very widely adopted. This has been largely due to local considerations, such as inadequate transportation facilities, lack of farm laborers, and the small amount of stock kept on farms, added to the undoubted attractiveness of the quick returns from growing wheat for a man with limited capital, who does not think enough of the rapid exhaustion of his soil from growing one kind of crop year after year. The increase of weeds, however, is Nature's protest against a one-year system. Nevertheless, at the present time great progress is noticeable in the West, in seeding down more land to hay and pasture and in increasing the number of cattle, horses and sheep, necessary to consume crops of a mixed nature. The following short rotation is recommended for the East by Mr. J. H. Grisdale, Agriculturist of the Central Experimental Farm:—"To destroy weeds, probably the best rotation possible is one of three years' duration including clover and mixed hay, followed by roots or corn, the land shallow-ploughed in fall and sown to grain the next spring with ten pounds of red clover and twelve pounds of timothy per acre. (When the land is heavy or clayey, the ten pounds of red clover may be replaced by six pounds of red clover and two of alsike.) If a portion of the arable land must be used for pasture, then the land might be allowed to remain under grass or hay for two years instead of one year, the second being used for pasture, thus extending the three-year into a four-year rotation. The pasture land in the four-year rotation, or the hay land in the three-year rotation, should be broken up early in August and cultivated at intervals to destroy the successive growths of weeds as they appear. The land should be again ploughed or preferably ridged in the fall. These rotations may be expected to give good results anywhere in Canada east of Manitoba."

Seeding Down.

The prevention of seed production by weeds is of the greatest importance when cleaning land for a crop. A useful way of choking out many perennial weeds and also of holding in check many annuals, is to seed down to grass or clover. This does not kill the seeds of all the different kinds of weeds which may be in the soil; but it prevents many weed seeds from being produced upon land which for one reason or another cannot be carefully worked. When the sod is broken up again, the seeds of some kinds will germinate and the plants become noticeable. In the same way that weeds crowd out crops and reduce the yield, so may weeds themselves be choked out by more vigorous plants which will prevent them from getting light and air, and thus weaklings are produced which bear only a few seeds instead of strong and vigorous plants which would produce perhaps many thousands, or even hundreds of thousands. Seeding down is always a wise practice on land which is not being used for some special crop, and is Nature's own plan for keeping up the fertility of the soil and preventing an undue preponderance of any one kind of plant.

Weeders and Harrows.

For the destruction of the seedlings of all kinds of weeds upon light land comparatively free of roots and stones, but which has been sown to grain crops for several years in succession, I know of nothing so effective as the working of the surface of fields of growing grain after the plants are three inches high, with the implements called Weeders, which deserve to be in much more general use than they are. In lieu of these, light harrows with sloped teeth may be used to advantage. The field should be dragged in the same direction as the drills. This work of course should only be done when the land is in the proper condition for harrowing. If fields are harrowed before the grain is three inches high, some of the plants are apt to be covered up too deeply; but after that stage of growth they are very much benefited by the operation, which not only kills all of the small weak seedlings of weeds which have germinated near the surface of the soil, without dragging out or in any way injuring the grain plants, but at the same time breaks up the surface of the soil. This benefits the crop so treated in exactly the same way as crops of corn or potatoes are helped by harrowing, as it forms on the surface a mulch of dry dust which holds in moisture. A single harrowing is generally sufficient; but, if necessary, the operation may be repeated two or three times and until the plants are six inches high. For such persistent annual weeds as Stinkweed, the various kinds of Mustard and Lamb's-quarters, the use of weeders or harrows as above recommended, is undoubtedly the most economical and in every way the best method of control.

Weeds not "Natural to the Soil."

It may not be amiss to refer to a widespread, although not always acknowledged, misconception on the part of some of our cultivators of the soil. Not a few people believe that weeds are what they term "Natural to the soil," by which they mean that these plants can arise in some way by spontaneous generation from the soil in places where they have not been sown and where there are no seeds. This is absolutely erroneous. No plant can ever begin to grow, except from a seed or from a piece of a similar plant in the ground. The appearance of weeds on land which has been cleaned, is due to the enormous number of seeds produced by some plants which have become weeds, and to the numerous ways in which these can be carried by the elements or otherwise transported to an adjoining piece of land, where, if the conditions of growth are favourable they flourish. The plants which have developed into noxious weeds, are those which have the greatest powers of propagating themselves and which are of robust habit and are better able to care for themselves than the plants we grow as crops, most of which are of exotic origin and of improved strains, such as are not able to hold their own unless cared for and helped by man. If plants could come into existence by spontaneous generation, the various kinds would be more evenly distributed; but, in the case of even our worst weeds, we know in many instances when and by what means they have been introduced into new districts, which would not be the case if they could spring up spontaneously from the soil.

Clean Seed.

Too much stress cannot be laid on the economy of using well cleaned seed for all crops, even if what seems to be a very high price has to be paid for it. It is undoubtedly the case that many weeds occur in crops, which can only be there from the seed having been sown with the crop seed. Examples of these are Chess in fall wheat, and Darnel and Cockle in spring wheat. The spread of weeds into new localities is almost entirely due to their having been introduced with seed brought from another district. The old idea that there was a great advantage in getting a change of seed from another farm or district, or from a crop grown on different soil, has been responsible for the bringing in of many troublesome weeds to farms where previously they were unknown. Many Canadian farmers are appreciating this, and more of them every year are adopting the wise plan of growing their own seed grain, carefully selected to a desired type, upon a plot of land specially prepared and kept clean for the purpose. The value of such work and the willingness on the part of farmers to pay a reasonable price, although apparently a high one, for good seed, clean, true to variety and type, and well matured, have given rise to that most useful organization, the Canadian Seed Growers' Association. The institution of such a plan of securing high class seed will affect the condition of the whole of a man's farm; for he will have an eloquent object lesson of what large yields may be produced by careful work on a small area, and also how much larger monetary returns may be secured in his whole crop by adopting similar measures. This has actually been the case with members of the Seed Growers' Association.

The Seeds of Weeds.

As so many weeds are introduced into new localities by their seeds being sown mixed with crop seeds, it is of the utmost importance that those who purchase seed from dealers should know how to recognize the seeds of the various kinds of these agricultural pests. Every farmer should know by sight the seeds of the fifteen or twenty worst weeds which are likely to be found among the seeds of cereals, clovers and grasses, as well as among some garden seeds. He should always examine for himself all crop seeds he buys, no matter what guarantee of purity he may get from the sellers. The seeds of most weeds are small and therefore it is necessary to examine them through a magnifying glass. A suitable glass, however, for this purpose can be purchased anywhere for twenty-five or fifty cents. All of the weed seeds have characteristic shapes, colours and markings, by which, after a little practice, they are just as easily recognized as the crop seeds among which they occur.

For ease of reference, it has been thought best to arrange the plants treated of in this bulletin in their natural botanical order, as given in Prof. Macoun's Catalogue of Canadian Plants. Where necessary, a short account has also been given of the different families of plants.

In the descriptive matter accompanying each plate will be found a careful description, giving the salient points by which the seed of each plant represented may be recognized; and at the end of the volume are given some plates showing these seeds, both of their natural size and also much enlarged so as to show the same seed as it appears under the ordinary pocket magnifying glass. In addition to the seeds of the plants figured and described, representations are also given of some other weed seeds which are likely to be found among crop seeds offered for sale in the market or which are likely to occur among crops grown by farmers in Canada. Each kind of seed on the plates is well represented in colours and is plainly marked with that name which is best known to the largest number of people interested, or which is used over the widest area of country. It is hoped that these will be useful to the large number of farmers, seedsmen and students, who are directing much more attention than formerly to this important subject.

Botanical Terms Explained.

In treating of the various weeds mentioned in this bulletin, it will be necessary for the sake of brevity to make occasional use of a few botanical terms which may not be familiar to all. A list of these is given below with explanations.

Achene—A dry one-seeded fruit in a hard close-fitting shell which opens only when burst by the germinating seed.

Anther—See Stamen.

Apical Scar—The mark on a fruit, where the style or stigma was attached.

Axil—The angle between a leaf and a stem.

Basal Scar—The mark on a fruit, where it was attached to the peduncle, or on a seed where it was attached to the seed vessel.

Bract—A small leaf bearing a flower in its axil.

Calyx—The outer set of leaves in a flower.

Compound—Composed of several similar parts.

Corolla—The inner set of leaves in a flower, generally coloured.

Corymb—A raceme in which the footstalks are gradually lengthened from the apex downwards, so that all the flowers are brought to the same level or nearly so.

Cyme—A panicle with the footstalks so developed or contracted as to form a flat-topped head, the central flowers generally blooming first; examples: Elder, Dogwood.

Dentate—With toothed edges.

Embryo—The rudimentary plant contained in a seed.
Entire—Not toothed.
Filament—See Stamen.
Fruit—The matured ovary and its contents, together with any appendages of the flower which seem to form an integral part, as the calyx of an Apple or a Rose.
Head—When numerous flowers are arranged upon a disk or receptacle; examples: Ox-eye Daisy, Clover.
Involucre—A circle of bracts round a flower or flower-head.
Irregular—With some of the parts different in size or shape.
Lobed—Divided to about the middle.
Ovary—See Pistil.
Panicle—A compound raceme, or a raceme with branched footstalks; example: Oats.
Pedicel—The stalk of a flower in a cluster.
Peduncle—The stalk of a flower.
Petals—The separate parts of a corolla.
Petiole—The stalk of a leaf.
Pinnate—Feather-like, having leaflets on each side of a main stalk.
Pinnatifid—Cut like a pinnate leaf.
Pistil—The female organ of a flower, composed of the *ovary*, which contains the seed, the *stigma*, a soft viscid part of the pistil which receives the pollen grains, and the *style*, which supports the stigma.
Pollen—See Stamen.
Pubescent—Downy.
Raceme—Like a spike but with the flowers borne upon footstalks of an equal and of a noticeable length; example: Lily-of-the-Valley.
Radicle—The first root that comes from a seed.
Regular—With the parts uniform in size and shape.
Rootstock—A creeping stem below the surface of the ground.
Runcinate—Having the teeth of a leaf directed towards the base.
Scarious—Membranous.
Seed—The embryo with its covering, if this is not part of the ovary.
Sepals—The separate parts of the calyx.
Sessile—Without a footstalk.
Spatulate—Expanded above and narrowed at the base.
Spike—When the flowerstalks are very short or wanting on a long cylindrical flower-cluster.
Stamen—The male organ of a flower composed of the *anther*, which holds the fertilizing *pollen* grains, and the support called the *filament*.
Stellate—Shaped like a star.
Stigma—See Pistil.
Stipule—A small leafy expansion of the base of a petiole.
Style—See Pistil.
Truncate—Cut off abruptly.
Umbel—When the flowers are supported upon footstalks rising from the summit of a general footstalk; example: Geranium. If each of the footstalks of an umbel bears a secondary umbel as in the Carrot, it is a *compound umbel*.

In Botany the word *fruit* signifies the enlarged and matured ovary, whatever its substance may be and whether fit to eat or not. In the small fruits of many weeds it is sometimes difficult to decide whether these are fruits or true seeds. In the Buttercup, Sunflower, Borage, and Mint families, the seed-like bodies are really fruits, while in the Mustard, Pink, Pea and Evening-primrose families they are true seeds. In describing the weeds in the present publication, it seems wise to speak of all of these as *seeds*, which is the term commonly used by seedsmen, farmers and others; but in the short notices of the Families of Plants the true nature is mentioned. Dr. L. H. Grindon, the eminent English botanist in his "British and Garden Botany" makes the following concise distinction: "There is an infallible distinction between a fruit and a seed, however much they may resemble each other:—the fruit always has a scar at the base, showing where it was attached to the peduncle, and *another* upon the summit, indicating the former presence of the style or stigma; but the seed has never more than *one* scar, indicating the point at which it was connected with the pod that contained it."

Certain of the worst weeds have been legislated against by the Dominion Parliament or by the provincial legislatures. These are all mentioned in this bulletin, and the province in which they have been proclaimed noxious is indicated under each species after the word Noxious as follows:—*Noxious*: Dom., Ont., Man., N. W., B. C.—meaning that laws have been passed by the Dominion and the provinces mentioned, looking to the destruction of the weed in the field or to the elimination of the seeds from crop seeds offered for sale.

Conclusion.

It should be remembered that all kinds of weeds can be kept under control on land worked properly and under a short rotation of crops, with the ordinary implements of the farm, used at regular time. Many of the recommendations made in this bulletin are special measures for cleaning land which has become badly infested by certain weeds.

THE BUTTERCUP FAMILY, *RANUNCULACEÆ*.

The Buttercup family contains a few weeds of secondary importance, as the Tall Buttercup, *Ranunculus acris*, L., which occurs in almost every part of the Dominion, and the Creeping Crowfoot, *Ranunculus repens*, L., which is troublesome in pastures in the Maritime Provinces, and is also found in other districts. Both are perennials which are acrid and irritating to the mouths of stock when eaten in a fresh state, but lose their noxious qualities when made into hay. The Cursed Crowfoot, *Ranunculus sceleratus*, L., is so acrid that the juice will blister the skin. The Sea-side Crowfoot, *Ranunculus Cymbalaria*, Pursh, has been suspected of poisoning stock in the North-west. The flowers in this family are either regular or irregular; the fruits are very variable, and the seeds may be contained in berries, in dry pods, or in achenes, which are small, separate, seed-like dry fruits, containing a single seed in a hard, close-fitting shell which does not open of itself, but remains closed until burst by the germinating seed. In the division of the family to which the true buttercups belong, the fruit is a head made up of several achenes. The seeds (achenes) of several kinds of Buttercup may be found in commercial grass seeds and are [Plate 55, fig. 42—natural size and enlarged 4 times] mostly flattened, somewhat oval in outline, pointed at one end, dark coloured and margined, from one-tweltfh to one-eighth of an inch across. None of the species of Butter-cup are deep-rooted, and all can be controlled by breaking up the pastures and re-seeding to grass. In this family there are also some virulently poisonous plants, as the Larkspurs (*Delphinium*) and Monkshoods (*Aconitum*) of the western plains. The Crocus Anemone, *Anemone patens*, L., var. *Nuttalliana*, Gray, has been the cause of losses in flocks of sheep. Owing to its earliness in flowering and its succulence, the hairy stems are eaten in quantity and the copious hairs, remaining undigested, form balls in the stomachs of the sheep. The White or Pennsylvanian Anemone, *Anemone canadensis*, L., sometimes increases so much in low pastures that it crowds out the grasses and necessitates the breaking up of the sod. The seed [Plate 55, fig. 41—natural size and enlarged 4 times] is sometimes found with the seeds of grasses. At figure 42, on the same plate, is the seed of the Tall Buttercup, a common impurity in the seeds of the coarser grasses. [1½ natural size and enlarged 4 times.]

THE FUMITORY FAMILY, *FUMARIACEÆ*.

An occasional weed in the wheat fields of Manitoba is the Golden Fumitory, *Corydalis aurea*, Willd. This plant sometimes appears in low land and in restricted areas, in such numbers as to crowd out grain crops sown on stubble. It is a succulent biennial with golden-yellow irregular flowers. the seeds [Plate 55, fig. 43—twice nat. size and enlarged 4 times] are shining black, rounded-kidney-shaped, about one-twelfth of an inch across, borne in one-celled, square, somewhat knotty, curved pods which split down one side to shed the seeds. Spring or fall plowing or the disking of stubbles before sowing will clear them of this weed.

THE MUSTARD FAMILY, *CRUCIFERÆ*.

There are few orders of plants of so great economic importance as the Mustard family, not only from the large number of troublesome weeds it contains, but also from the fact that not a single unwholesome plant is found in it. Many species form well known and excellent articles of food, as the cabbage, turnip, radish, watercress, etc. Some of our worst weeds, however, most of which have been introduced into Canada, belong to this family. The characters of the family are easily recognized. The flowers are regular, composed of four free sepals and four free petals, arranged in two opposite pairs and forming a cross-shaped flower, from which the whole family takes its name, *Cruciferæ*. The flowers are borne on footstalks (pedicels), and clustered at the tips of branches which gradually elongate, forming long, upright racemes, with, often, fully-formed and even ripe pods below, before the topmost flowers have opened. When ripe the seed pods, which are of various shapes, in nearly all instances, consist of two outside walls separated by a thin white partition, to the two sides of which the seeds are attached. The seeds are, as a rule, small and very numerous, Their surface is usually rather dull, more or less granular, and many, when placed in water, develop gelatinous hairs and mucilage, by which, when they dry, they remain attached to passing objects or adjacent surfaces. This constitutes an important factor in their distribution. The quantity of mucilage varies. On some seeds it remains after drying as a covering of hair-like points or threads. The seed-coat is generally thin and close-fitting, the shape of the embryo plant showing plainly through it. The position taken by the folded up embryo inside the seed is often a great help in identifying the seeds of weeds. The seed leaves and radicle inside the seeds of crucifers take one of four characteristic positions:

1. Accumbent, when the radicle lies along the edges of the seed leaves, as in Stinkweed, from a Latin word *accumbo*, meaning to lie at the side.

2. Incumbent, when the radicle lies down the back of one of the seed leaves, as in Shepherd's-purse, Common Pepper-grass and Hare's-ear Mustard, from a Latin word *incumbo*, meaning to lie on the back.

3. Oblique, when the radicle lies slantingly across the edges of the seed leaves, as in the Prairie Wallflowers.

4. Conduplicate, when the radicle lies on the back of one of the seed leaves and these are folded sideways over it, as in the Mustard, Radish, etc.

Many of the Crucifers have in the roots, stems or seeds recognizable odours or flavours, which help in identifying them. The leaves and stems of many bear small star-shaped hairs. The best known or most noxious members of the Mustard family are treated of separately and figured, others which might be confounded with them will be referred to in the text, and the chief differences pointed out.

Closely allied with the Mustard family is the Caper family (*Capparidaceæ*), which has some important characters in common with it, such as cross-shaped flowers, seeds in pods, but these without partitions, and often having acrid or pungent juice. The chief differences between the two families are that in the Mustard family, four of the six stamens are long and two short, while in the Caper family all six are equal, and the pod of the former is two-celled, being divided by a thin partition across the middle. The curious seeds [Plate 55, fig. 47—natural size and enlarged 4 times] of the ENTIRE-LEAVED SPIDER-FLOWER, *Cleome integrifolia*, T. & G., of this family are sometimes found in western wheat. These seeds are rounded-wedge-shape or elongated-kidney-shape, with a deep curved groove running up each face two-thirds of the way to the top from just above the sharp-pointed base. These seeds, when ripe, are dark brown, roughened with pale, scurfy excrescences; the dried unripe seeds are yellowish.

Plate I

WORM-SEED MUSTARD
(*Erysimum cheiranthoides*, L.)

PLATE 1.

WORMSEED MUSTARD, *Erysimum cheiranthoides*, L.

Other English name: Treacle Mustard.

(Noxious: N.W.)

Native. Annual and winter annual. Stems erect, simple or branching, six inches to two feet high. Dark green; whole plant sometimes slightly hoary with very short star-like hairs. Leaves, lance-shaped, sparsely toothed. Flowers bright yellow, one-fifth inch across, in terminal clusters about one inch across, on gradually elongating racemes. Seed pods slightly curved from half an inch to one inch long, obtusely four-angled, erect on spreading pedicels. Each pod contains about twenty-five seeds. An average plant will ripen about 25,000 seeds. Seeds [Plate 53, fig. 1—twice nat. size and enlarged 8 times], variable in size and shape, many being pointed at one end, rounded at the other, about $\frac{1}{24}$ of an inch long, reddish yellow, with a dull surface, but almost destitute of mucilage. Scar end darkened. Radicle conspicuous and incumbent, that is lying down the back of one of the seed leaves, which are plainly recognizable in the dry seed. The taste of the seeds is very bitter, which renders them unpalatable to stock, and it is claimed by some feeders that they are very injurious. This contention, however, is not borne out by recent experiments at Port Arthur, where large quantities have been fed to sheep without any injury.

Time of Flowering: June to autumn; seeds ripe July to frost.

Propagation: By seeds.

Occurrence: Frequent in waste places and on cultivated land throughout Canada.

Injury: A common impurity in clover seed, also a weed of summer-fallows and grain crops grown on stubble, occasionally so abundant as to crowd out grain.

Remedy: The injuries by Wormseed Mustard are mostly by those plants of which the seeds germinate in autumn, and which remain on the land through the winter. The destruction of these, however, is a simple matter by ploughing the land in fall or spring. A disk-harrow may be used with good effect if the work is done early in the spring before too many new roots are made.

PLATE 2.

HARE'S-EAR MUSTARD, *Conringia orientalis* (L.) Andrz.

Other English names: Rabbit-ear, Hare's-ear Cabbage, Klinkweed.

Other Latin names: *Erysimum orientale,* R. Br.; *Brassica orientalis,* L.; *Brassica perfoliata,* Lam.; *Conringia orientalis,* Reich.

(Noxious: Dom., Man., N.W.)

Annual and winter annual. Introduced from Europe, probably with flax seed, about 1892, now general throughout Manitoba and the North-west. Stems erect, with a few branches, 1 to 4 feet high. Whole plant perfectly smooth and glaucous (grayish green), with the appearance of a cabbage, when young. Leaves fleshy, entire, near the root obovate, gradually narrowed to the base; on the stiff stems, which become very wire-like when ripe, oblong oval, shaped like a hare's or a rabbit's ear, clasping the stem by two rounded auricles. Flowers creamy white, $\frac{1}{4}$ inch across. Pods square, 3 to 4 inches long. Seeds [Plate 55, fig. 44—twice nat. size and enlarged 4 times] dark brown, rounded oblong, pointed at the scar end, $\frac{1}{12}$ of an inch long, granular-roughened; when soaked in water, covered with a thick pile of short erect white mucilaginous hairs.

Time of Flowering: End of June; seeds ripe August to September.

Propagation: By seeds.

Occurrence: In grain fields, on stubble and by roadsides wherever grain is carried.

Injury: This succulent plant absorbs much moisture from the soil, very little grain growing where there is a patch of it. The wiry stems are hard to bind and an infested crop requires more labour to handle and much more twine to bind it.

Remedy: Pull by hand. If in large quantity, summer fallow. Disk stubbles in fall or early spring to kill plants which germinate after harvest and live over winter.

Plate 2

HARE'S-EAR MUSTARD
(Conringia orientalis (L.) Andrz.)

Plate 3

GREEN TANSY MUSTARD

(Sisymbrium incisum, Engelm., v. filipes, Gray.)

PLATE 3.

GREEN TANSY-MUSTARD, *Sisymbrium incisum,* Engelm., var. *filipes,* Gray.

Other English name: Cut-leaved Tansy-Mustard.
Other Latin names: *Sophia incisa* (Engelm.) Greene.

(Noxious: N.W.)

Native. Biennial. In the first season a rosette of finely divided leaves lying on the ground. Stems, 3 to 4 feet, erect, widely branching at the top and bearing an enormous number of narrow, smooth, slightly curved pods from $\frac{1}{2}$ to $\frac{3}{4}$ inch long on slender spreading pedicels. Whole plant bright green and somewhat glandular. Leaves pinnately divided and the pinnæ again once to twice divided into linear-oblong entire or toothed segments. Flowers yellow, one-eighth of an inch across in elongated racemes. Seeds [Plate 53, fig. 2—natural size and enlarged 8 times] very small, $\frac{1}{25}$ of an inch long, reddish brown, minutely roughened with mucilaginous hairs, shedding out very easily.

Time of Flowering: July; seeds ripe August.

Propagation: By seeds.

Occurrence: In crops grown on stubble and on imperfectly cultivated summer-fallows.

Injury: A coarse unsightly weed and a gross feeder.

Remedy: Fall and spring cultivation and hand-pulling.

The GRAY TANSY-MUSTARD, *Sisymbrium incisum,* Engelm., var. *Hartwegianum,* Watson, is also a tall coarse biennial plant with much divided foliage like the above, but differs by being covered with short gray pubescence and its more erect habit of growth. It has pods only $\frac{1}{4}$ inch long, all crowded close to the slender ascending branches which form a narrow spire. The Gray Tansy-Mustard is the commoner and more widely distributed plant of the two. It flowers and ripens its seed some weeks later. These two coarse biennials grow only from seed, but they throw out long branches from their white tap roots and draw nourishment from a wide area. As they stand up considerably above the crop, they are a conspicuous advertisement of negligent farming.

PLATE 4.

TUMBLING MUSTARD, *Sisymbrium altissimum*, L.

Other English name: Tall Sisymbrium.

Other Latin names: *Sisymbrium sinapistrum*, Crantz; *Sisymbrium pannonicum*, Jacq.

(Noxious: Dom., Man., N.W.)

Annual and sometimes winter annual. Introduced into the Prairie Provinces from Central and Southern Europe about 1887. Two to six feet high; stem branching, the lower part and the root leaves downy and glandular, with a musky odour. Upper part of the stem and the much divided leaves smooth. The young plants form a rosette of soft pale green downy leaves, shaped much like those of the Dandelion. On the flowering plants the leaves change very much in shape from the root up, no two being alike. Flowers pale yellow, $\frac{1}{3}$ inch in diameter, cross-shaped as in all the members of the Mustard and Cress family. Seed pods 2 to 4 inches long, very slender and produced abundantly along the branches. Each pod contains about 120 seeds, and a single plant has borne one million and a half seeds. Seeds [Plate 53, fig. 3—natural size and enlarged 8 times] very small $\frac{1}{25}$ of an inch, olive-brown or greenish-yellow, minutely roughened with mucilaginous glands, oblong, angular, truncate at the scar end, sometimes almost square from compression in the pod, grooves between the edges of the seed leaves and between these latter and the radicle conspicuously darkened. The seed leaves and incumbent radicle plainly visible through the thin skin. When the seeds are ripe the whole head of the plant breaks off and is blown across the prairie, scattering the seeds far and wide. The seeds, as in many "tumbling weeds," are not very easily shed from the tough pods, consequently a head of this weed may blow about on the prairie for a whole winter, dropping a few seeds at intervals for many miles.

Time of Flowering: June to July; seed ripe August.

Propagation: By seeds.

Occurrence: In grain fields in the West. Occasionally found along railways and in waste places in other parts of Canada, but not as a farm weed.

Injury: This is a Mustard with all the bad characteristics of those aggressive enemies of the farmer—enormously prolific, with great powers to spread, owing to its tumbling habit; a coarse, conspicuous plant and a gross feeder. The seed however is so small that with a little care it can be easily cleaned from seed grain. It does not appear to retain its vitality in the soil as long as the seed of some other kinds of Mustard.

Remedy: Hand-pull when there are only a few plants. Pay particular attention to edges of fields and fire-breaks. Cultivate growing crops with weeder or light harrows.

TUMBLING MUSTARD
(*Sisymbrium altissimum*, *L.*)

Plate 5

WILD MUSTARD OR CHARLOCK
(Brassica Sinapistrum, *Boiss.*)

PLATE 5.

WILD MUSTARD, *Brassica Sinapistrum*, Boiss.

Other English names: Charlock, Herrick, Cadluck, Field-Kale, Ontario Mustard.

Other Latin names: *Sinapis arvensis*, L., et auct. plur.; *Brassica Sinapistrum*, Benth.

(Noxious: Dom., Man., N.W.)

Annual. Introduced from Europe. Now found in all parts of Canada, but most abundant in the Eastern Provinces and Manitoba. The erect branching stems 1 to 3 feet high, rough with stiff, somewhat deflexed, hairs. Lower leaves stalked, usually deeply indented or lobed, with the terminal lobe large. Upper leaves mostly sessile. Flowers bright yellow, fragrant, ⅔ inch broad. Seed pods 1 to 2 inches long, knotty or slightly constricted between the seeds, ribbed and ascending, on short thick pedicles, tipped with a long empty or 1-seeded 2-edged beak, which comes away whole from the ripe pod. Seeds [Plate 53, fig. 4—natural size and enlarged 8 times] round, about 15 to 17 in each pod, very dark brown or reddish black and almost smooth, no mucilage. The seeds have great vitality; an actual proved instance of their having lain in the ground in a salt marsh in Nova Scotia for twenty years is known to me, and another of Wild Mustard appearing abundantly on the ploughing up of a pasture which had been down for twenty-four years. A purple patch at the junction of the branches with the stem is a striking character of this Mustard.

Time of Flowering: June to September; seed ripe by August.

Propagation: By seeds only.

Occurrence: In all farm crops and in waste places. Distributed in crop seeds, by floods and wind, and in manure.

Injury: A gross feeder and recognized generally as an indication of negligent farming.

Remedy: Hand-pull regularly if only in small numbers. Harrow stubbles as soon as the crop is harvested to start autumn growth; cultivate down the first growth or feed off with sheep. Leave late plants, which the frosts of winter will kill before seeds ripen. Spraying young Mustard plants with a 2 per cent. solution of bluestone (sulphate of copper), that is, 2 lbs. of bluestone in every 100 lbs. (10 gallons) of water, has given very satisfactory results. The bluestone solution falling on the leaves and tender stems of the Mustard kills them in a few hours without any injury to the grain or grass crop amongst which they are growing. The work must be done when the Mustard is quite young and succulent, that is, when the first flowers are opening. A barrel, 40 gallons, will almost cover an acre if used with care and put on at the best time, before the plants are too large. The cost is about 80 cents per acre. This method is a practical one in Ontario, where water is plentiful and labour and implements are cheap. In the West this is not always the case, and a far better plan on the large prairie farms where seeding down with a grain crop is seldom resorted to, is to kill the young plants of Wild Mustard and all other annual weeds, the seeds of which germinate near the surface of the soil, by harrowing growing crops with weeders or light harrows after the crop is well up and there is no danger of covering the blade too deeply. Some have hesitated to adopt this method, fearing that the grain plants would be dragged out by the harrows. No such fears need be entertained, if the work is done when the land is in the proper condition for harrowing and with light, slope-toothed harrows. It is suggested that all who have trouble with annual weeds, such as the various kinds of Mustard, Stinkweed, Ragweed, Lamb's-quarters, Wild Buckwheat, etc., should at any rate try this method on a small part of their crop; one strip up the side of one field will be enough to show the good results to be obtained. In addition to the real Wild Mustard there are two or three other troublesome species which may be confounded with it, and are all to be regarded in the same light by farmers on account of their aggressiveness.

Bird Rape, or German Rape, *Brassica campestris*, L. (including *Brassica Rapa*, L., and *B. Napus*, L., which cannot be separated by any permanent characters). This resembles the Wild Mustard very closely in general appearance, but has only the root leaves hairy. The stem, upper leaves which clasp the stem by an auricled base, and the long pods on spreading pedicels, are all perfectly smooth and waxy like the leaf of a cabbage. This form is abundant in Manitoba and in some parts of the Provinces of Quebec and Ontario. An important difference between Bird Rape and Wild Mustard is that the former cannot be killed by spraying with the bluestone solution.

Indian Mustard, *Brassica juncea*, L., which closely resembles Bird Rape, has been detected in a few places in Manitoba and Ontario. It differs chiefly in the stem leaves, which have not clasping bases, in the shape of the pods and in the shorter pedicels, which are less spreading.

Black Mustard, *Brassica nigra* (L.) Koch. Although appearing occasionally in all of the Provinces and often mentioned by correspondents, the true Black Mustard does not, as far as I am aware, occur anywhere in Canada as a troublesome weed upon farms. It may be at once known by its long spreading branches covered with short square pods only half an inch long, which are erect and closely appressed to the stem.

Wild Radish, *Raphanus Raphanistrum*, L. Much of the "Wild Mustard," "Cadluck" or "Kale" of Nova Scotian correspondents is really the Wild Radish, which is an annual 1 to 2 feet high with a few long branches starting low down. The root is slender, not swollen as in the Garden Radish. Leaves pale yellowish green, deeply lobed and bearing like the stem a few stiff bristles. The flowers are fewer and larger than in Wild Mustard, noticeably paler yellow, conspicuously veined. The constricted seed pods give the best characters, and with these no mistake can be made between these two similar plants. In Wild Radish the seed-pods have no valves but are composed of two joints, the lower one small, one-tenth of an inch, and seedless, which remains attached to the footstalk: the upper, cylindrical, 1½ inches long, with several one-seeded cells formed by transverse partitions. This seed-bearing portion separates from the first joint, leaving it attached to the pedicel and in threshing is broken up into single-seeded sections.

All of the above weeds may be treated in the same way, except that the Wild Radish, like the Bird Rape, cannot be destroyed easily by spraying with the bluestone solution.

Plate 6

FALSE FLAX
(Camelina sativa, Crantz)

PLATE 6.

FALSE-FLAX, *Camelina sativa*, Crantz.

Other English name: Gold of Pleasure.

Other Latin names: *Myagrum sativum*, L.; *Camelina sativa*, Fries.; *Camelina macrocarpa*, Reich.

(Noxious: Dom., Man., N.W.)

Introduced from Europe. Annual and winter annual, 2 to 3 feet high; erect, branched above. Root leaves lanceolate and narrowed into a petiole; upper leaves arrow-shaped, sharply pointed. The lower leaves and the lower part of the stem downy with star-shaped hairs. The upper part of the stem smooth and glaucous. Flowers numerous, small, one-eighth of an inch across, pale greenish yellow. Pods three-eighths of an inch, balloon-shaped or pear-shaped, margined and tipped with a slender beak, on slender ascending pedicels, each containing about 10 seeds. Racemes much elongated in fruit. Seeds [Plate 55, fig. 45—natural size and enlarged 4 times] very variable in size, about $\frac{1}{20}$ of an inch, pale yellowish brown. The groove between the radicle and the seed leaves distinct, that between the seed leaves less so. Seed coat finely pitted. Scar of attachment white, in a notch between the tip of the radicle and the seed leaves. When the seeds are soaked in water, they develop a copious coat of mucilage and a fine pile of transparent hairs. The seeds have been used as food on account of their mucilage and oil, both of which resemble those of linseed. The plant has been considerably cultivated for these products in Germany and France.

Time of Flowering: June to August; seed ripe July to September.

Propagation: By seeds only.

Occurrence: All through Canada in waste places and along railways. A noxious weed in the West and in western Ontario.

Injury: A common weed in the West in grain crops, particularly on stubble, and in flax fields. In Ontario, in fields of fall grain. The seeds are frequently found in the seed of flax, clover and grass.

Remedy: Hand-pull, surface cultivation in fall and spring. Summer-fallow early.

In Ontario this and similar weeds may be kept in check by summer-fallowing early and then sowing the same season with rape in drills 26 inches apart, the horse hoeing to be supplemented with more or less hand hoeing. This may be followed the next spring by another hoed crop as corn, potatoes or turnips. Rape is one of the best cleaning crops that can be grown if carefully cultivated, and where it is followed the next year with a hoed crop, few weeds can withstand the treatment.

There is another species of False-flax which is found occasionally, and which is rapidly becoming more abundant and wide spread, both in the West and in Ontario. This is the SMALL-SEEDED FALSE-FLAX, *Camelina microcarpa*, Andrz., which resembles the ordinary weed very much, but has smaller pods and smaller and darker-coloured seeds.

PLATE 7.

BALL MUSTARD, *Neslia paniculata* (L.) Desv.

Other English names: Yellow-weed, Neslia.

Other Latin name: *Myagrum paniculatum*, L.

(Noxious: Dom., N.W.)

Introduced into the West about the same time as Tumbling Mustard, Hare's-ear Mustard and Cow Cockle. A tall slender, annual or winter annual, which has spread through the grain growing districts on the prairies with great rapidity, until it is now found as a bad pest of the grain grower from Manitoba to the Pacific. Stems erect, very slender; strong plants throwing out a few long branches. Whole plant yellowish green and covered with small appressed star-shaped hairs. Lower leaves lance-shaped, narrowed at the base; stem leaves arrow-shaped, clasping the stem at the base, blunt-pointed. Flowers small, $\frac{1}{8}$ of an inch, orange-yellow; racemes very long, with the small, round, one-seeded, shot-like pods [Plate 55, fig. 46—natural size and enlarged 4 times] standing out from them in all directions on slender footstalks about half an inch in length. The pods do not open to discharge the seed, but dry up and produce a small, roundish, brown, wrinkled object, like a small piece of dry earth, about $\frac{1}{12}$ of an inch across. The contained seed is yellow.

Time of Flowering: June to August; seed ripe July to September.

Propagation: By seed only.

Occurrence: In grain fields all through the West. In the East along railways and wherever western grain is carried. Ball Mustard is troublesome as a weed only in the West.

Injury: This weed has spread chiefly from the inconspicuous nature of the seed. It is frequently overlooked in seed grain, owing to the resemblance of the persistent wrinkled pod to a small particle of earth.

Remedy: Early summer-fallowing and the disking of stubbles in fall and spring are perhaps the best way to hold it in check. All seed grain should be very carefully cleaned before sowing. In very badly infested fields the stubble should be harrowed as soon as the crop is harvested to start a crop of seedlings, which should be disked down late in autumn. The next spring the land may again be cultivated and sown late to early barley, which should be cut on the green side, or oats may be sown for green feed. The edges of fields should be mown before the seeds of the Ball Mustard are ripe, and the hay fed at once or burnt. This and all the mustards make good green feed.

Plate 7

BALL MUSTARD
(Neslia paniculata, (L.) Desv.)

Plate 8

SHEPHERD'S PURSE
(Capsella Bursa-pastoris, *Moench.*)

PLATE 8.

SHEPHERD'S-PURSE, *Capsella Bursa-pastoris,* Moench.

Other Latin names: *Bursa pastoris,* Weber; *Thlaspi Bursa-pastoris,* L.

(Noxious: N.W.)

Introduced. Annual and winter annual. The Shepherd's-purse is well known to every gardener. At all times of the year, when it is not actually freezing, this plant is growing and ripening its seeds.

There are few plants so variable in size and appearance. There may be at the base a vigorous rosette of leaves, or there may be none at all. The leaves may be deeply cut, pinnatifid, or quite entire, without any teeth at all. The plant may be bright green and nearly smooth or canescently gray with short hairs and stellate down. A seed-bearing plant may be a dwarf, little more than an inch or two high, or a vigorous branching plant three feet high with many pods. The stem leaves are for the most part arrow-shaped, with two sharp ear-like projections, one on each side of the stem. The flowers are small and white. The only part of the plant which seems to be constant is the seed pod, which is flat, triangular in shape, $\frac{1}{4}$ of an inch long, wedge-shaped at the base, notched at the top with the outer angles rounded. Each pod contains about 20 seeds. These [Plate 53, fig. 5—natural* size and enlarged 4 times] are small, $\frac{1}{24}$ of an inch, oblong, reddish brown, the surface dull and punctured. When put in water, they produce a large amount of mucilage and a covering of rather long but very fine transparent hairs.

Time of Flowering: Throughout the season.

Occurrence: Throughout Canada, in all soils.

Propagation: By seed only.

Injury: The injuries due to this prolific weed are not, I think, fully appreciated by farmers. It has an enormous power of propagation, a single plant will ripen 50,000 seeds. It spreads rapidly and sometimes takes entire possession of land. It will thrive in all kinds of soil; is of a succulent nature and absorbs a large amount of moisture from the soil. In meadows which have been thinned by winter, the vigorous rosettes of autumn-started plants will crowd out grass and clover. The seed is found in all small commercial seeds. Mr. Willing reports it as doing serious damage in wheat fields in the North-west.

Remedy: Although so aggressive and persistent, this weed can be easily destroyed if farmers understand how much injury it may do them. The small central root is easily cut by hoe or cultivator, and the plant is not remarkable for vitality. Constant hoeing in gardens and early summer-fallowing with frequent cultivation will easily clear infested farm lands.

* The figure to show the natural size is slightly too large.

PLATE 9.

STINKWEED, *Thlaspi arvense*, L.

Other English names: Penny-cress, French Weed.

(Noxious: Dom., Man., N.W.)

Introduced. Annual and winter annual, with an abominable smell when bruised. The most persistent and aggressive enemy of the western wheat grower. Plants in bloom when winter sets in freeze up; but, as soon as they thaw out in the spring, they continue to grow and mature their seed without the slightest injury. The seeds of these early plants are ripe early in July. Plants which grow from seed in the spring are not ripe until some weeks later. Erect. stem simple or branching. Whole plant bright green and quite smooth. Root leaves petioled; stem leaves spear-shaped, coarsely toothed, clasping the stem at the arrow-shaped base. Flowers clear white, one-eighth of an inch across. At first in a small flat cluster at the top of the leafy stem. Racemes elongated in fruit. Pods flat, $\frac{3}{4}$ of an inch across, containing from 8 to 16 seeds, on slender upward-curved pedicels, pale green and winged, notched at the top. Just before the seeds ripen, the pods turn to a characteristic greenish orange shade, which is easily noticed in fields when this weed is growing among crops. Seeds [Plate 53, fig. 6—natural size and enlarged 8 times] deep purplish brown, asymmetrically oval in outline, flattened, with rounded edges, about $\frac{1}{12}$ of an inch across, a little longer than broad. The surface of the seed has on each face five or six loop-like lines, which start at the basal scar or notch and run concentrically around a central groove, which starts between the radicle and the seed leaves. The seed is a beautiful object when examined under a magnifying glass. It is also interesting from having the seed leaves, inside the seed, flat and with the radicle lying along their edges on one side (accumbent). There is no mucilage developed on these seeds when soaked in water.

Time of Flowering: Throughout the season; seed ripe from July to frost, and after the middle of June too far advanced to be ploughed down without danger.

Propagation: By seeds.

Occurrence: Stinkweed is now found in every province of Canada, but it is nowhere such a terrible pest to the farmer as in Manitoba and the North-west. It was introduced on to the prairies at the time of the first settlers, and its worst occurrences to-day are along the lines of the old trails; but it is rapidly spreading into new districts.

Injury: On account of the great loss due to the presence on land of this noxious weed, it is of the utmost importance that everyone should know its appearance, so as to destroy it whenever it shows itself in a new locality. The North-west Government, through their Chief Weed Inspector, Mr. T. N. Willing, have adopted the wise plan of publishing an accurate and easily recognized coloured picture of this weed. This was not only issued in their Bulletin No. 16, "Hints for the Grain Grower," by Mr. Willing; but, in addition, a copy of it was put in every school in the Territories so that teachers and students might become familiar with the appearance of

Plate 9

STINKWEED OR PENNYCRESS
(Thlaspi arvense L.)

Stinkweed. This pest is found in the greatest abundance in the rich lands of the Red River valley, where it flourishes with remarkable vigour and where great difficulty is experienced in controlling it, on account of the distribution of the seeds by spring floods. There are also in other parts of the Province and in the North-western Provinces and British Columbia, certain districts where this weed has established itself thoroughly and where it is causing great loss.

Remedy: Hand-pulling and burning is probably the best way of stamping out Stinkweed in new localities and on small areas; but, when it is thoroughly established, more drastic measures must be adopted. These are all based on some operation by which the seeds are covered up so as to get them to germinate, and then the young plants are destroyed with harrow, cultivator or plough, before they ripen seeds. The land should be cultivated persistently as soon as a fresh growth of the weed has developed, and great care must be taken not to plough down any full-sized seed pods, even although they may be green, as it has been proved that seeds in such a state of development, can ripen beneath the soil in the dry climate of the West. On land which is to be summer-fallowed, if there is a heavy growth of this weed with fully formed seed pods, the plants should be first mowed down, removed from the field and piled on a piece of hard land where, after drying, they can be burnt. Mr. Willing says truly: "Close attention should be given to any portion of a farm where Stinkweed has been noticed, and careful persistent work will be required to eradicate it; but it should not be forgotten that it will pay well to drop all other work and fight this weed when it is first noticed." If a crop on new ploughing or summer fallow is noticed, early in the season, to be infested, the field should be thoroughly worked on the surface with a weeder or a light harrow with sloping teeth, as described under Wild Mustard at page 24, beginning the work when the grain is three inches high and repeating the operation once or twice afterwards in bad cases. It will be found that this destroys an enormous number of small weed seedlings and improves the grain crop very much. It is, I feel confident, the most important remedial measure which can be adopted to clear land of Stinkweed. Experiments of recent years in the Red River valley have shown this to be the case; and the remarkable sight may there be seen, of clean fields giving crops of forty bushels to the acre, with others close to them bearing not more than from ten to twelve bushels of grain, the difference being entirely due to the presence of Stinkweed in the one case and to its having been harrowed out in the other. A treatment which has given good results is to run a disk or harrow over stubble as soon as the crop is carried, so as to start into growth the seeds near the surface. The following spring, harrow or cultivate these plants down; and, as soon as a growth of fresh plants starts, plough the land and harrow at once. This land may be sown late to a green feed crop, or it may be kept under a clean fallow for the whole season if the land can be spared. The following spring any growth of weeds should be cultivated down before sowing the crop. An excellent plan which will enable a farmer to take a rest in this fight, so as to attend to other parts of his farm, is to seed down with Brome Grass or Western Rye-grass, fifteen pounds to the acre, in the western provinces, or to Timothy or Western Rye-grass in Manitoba. Such fields will require mowing occasionally during the first season to prevent seeds ripening. The grass will choke out even the Stinkweed after the first year. When the sod is ploughed up again, some seeds will germinate; but these can be dealt with, as before. The seeds of Stinkweed are very easily carried on threshing machines and other implements, so these should be carefully examined when coming from a district infested with Stinkweed on to a clean farm. Where only a small patch occurs on a farm, the greatest care must be exercised in treating it, if there are any ripe seeds, for these may be carried easily in mud attached to a boot or to the feet of horses and other animals, as well as on the wheels of vehicles.

The use of the name Stinkweed is recommended for this plant in preference to all others sometimes used. It is one of the worst weed enemies a farmer in the West has to fight. The use of this objectionable name does much to identify it, and there is no doubt that it has spread widely from not being recognized when it first appeared in new districts. The name "French Weed," very much used in Manitoba, is a senseless name for such a persistent enemy, because it tells nothing about the plant, and it is most doubtful whether the early French settlers in Manitoba were the means of its introduction.

S-t-i-n-k is a small word with a big meaning, easily understood by anyone who has ever handled Stinkweed, or tasted milk or butter from a cow which has eaten it.

Stinkweed is so called on account of its abominable smell.

Plate 10

PEPPERGRASS
(Lepidium apetalum, Willd.)

PLATE 10.

PEPPERGRASS, *Lepidium apetalum*, Willd.

Other Latin name: *Lepidium intermedium*, Gray.

Native. Annual and winter annual. Stems erect, profusely branching above, 6 inches to 2 feet high, somewhat hoary, with short appressed hairs. Autumn plants produce a rosette of dark green deeply indented leaves, much like some specimens of Shepherd's-purse, but more succulent. Stem leaves with a few coarse teeth, narrowed at the base. The many ascending and spreading branches give this plant, when in seed, the appearance of a miniature tree, the very numerous small nearly round flat pods taking the place of leaves; the real leaves of the plant fall away early when the seeds begin to ripen. Although there are only two seeds in each pod, a very large amount of seed is produced by each plant. The flowers are minute, and the seed pods about $\frac{1}{10}$ of an inch wide, heart-shaped, slightly longer than wide, notched at the tip; these separate into two valves in the same way as in the Stinkweed; but in this plant there is only a single seed in each valve. The seeds [Plate 53, fig. 7—natural size, and enlarged 8 times] are egg-shaped in outline, much flattened, blunt on one edge, and very thin at the other, where there is a more or less apparent narrow wing; each flattened side shows a rather deep groove between the radicle and the seed leaves, extending three parts of the way up the centre of the seed from the base. The basal scar of attachment is white, and projects somewhat from the outline of the seed. The colour of the seed coat is bright reddish-yellow. The wetted seeds give off a large supply of mucilage, the transparent hairs are of medium length and fewer in number than in Shepherd's-purse and False-flax. The embryo is incumbent, the radicle lying down the back of one of the seed-leaves, which is the easiest character by which this species and its close allies may be distinguished from the very similar *Lepidium virginicum*. This latter species, however, is very rare in Canada, and nowhere occurs as a weed.

Time of Flowering: June, July, seed ripe on early plants by the end of June.

Propagation: By seeds.

Occurrence: Throughout the Dominion, but seldom complained of as a weed in the East, except in clover in the clover-seed-growing districts of Ontario, and in grain crops of the West, where it crowds out crops grown on stubble, particularly on light land and in a wet spring.

Injury: Peppergrass has frequently been the cause of considerable loss in the West under the conditions mentioned above. This is from the autumn grown plants which get a start before the grain crop and choke it out with their thick vigorous growth. The seed is a frequent impurity in grass and clover seed.

Remedy: Clearing land of Peppergrass is an easy matter, because the plants which do harm are those which germinate in the autumn. Disking land in autumn or early spring, before the new growth of roots takes place, will clear it entirely of th's weed. If the work is not done early in spring the deep tap roots send out many side branches, and it will be necessary to use a duck-foot cultivator to cut the plants off or to give the land a light spring ploughing. On land which is to be summer-fallowed, patches of this weed should be mowed before ploughing; or, unless the work is done very early, many seeds will be turned down in a condition to grow. Summer-fallowing, from almost all considerations, should be done as early in spring as other farm work will permit of; the best time is directly after seeding is finished.

When clover is to be cut for seed, it will pay to mow with a scythe or pull by hand any plants of Peppergrass seen on bare spots.

FIELD PEPPERGRASS OR COW CRESS, *Lepidium campestre*, R.Br., occurs in a few places in western Ontario, and the seeds have been found from time to time in clover seed in the market. They resemble very much in a superficial way the seeds of Hare's-ear Mustard. They are about the same size and colour, but are of a redder shade of brown, and are of a rather different shape. When examined they can be easily recognized. The Field Peppergrass has seeds one-twelfth of an inch long, egg-shaped, but pointed at the scar end. The surface is finely roughened and dull, with apparently a mealy surface. This is due to the copious coat of mucilage, mixed with long gelatinous hairs, which is developed when the seeds are soaked in water. The space between the radicle and seed leaves is shown by a deep but narrow groove with sharp edges, extending from the sharp basal end almost to the top of the seed. The plant grows with two or three stems from the same root. These are corymbosely branched above, with a few ascending branches. The lower leaves are oblong, and not toothed. Those of the stem are spear-shaped with blunt ends, a few large shallow teeth and an arrow-shaped base. The thick seed pods are broadly ovate, boat-shaped, being rounded below, and hollowed out above. They are prominently keeled below, and are roughened with short fleshy prominences; they stand out stiffly from the stem on pedicels of about their own length. Each pod contains two seeds. The plant is a biennial, and is as yet rare in Canada, occurring only, as far as evidenced by the seeds, in the clover-growing districts of Ontario, and at one or two other points in the eastern counties of the same Province.

THE PINK FAMILY, *CARYOPHYLLACEÆ*.

The Pink family, which contains several beautiful garden flowers, also embraces some troublesome weeds, the seeds of which are frequently found in clover and grass seeds. The characters are well marked. All the weeds are herbs with brittle stems, articulated and thickened at the joints, frequently forked. Leaves entire, generally opposite and joining round the stem at the base. The flowers usually borne in corymbs, regular, of 4 or 5 parts, or double those numbers, with or without petals; stamens nearly always normally 10; by deficiency, sometimes only 5, 4, or 3; styles 2 to 5, rarely united into one. Seeds usually many, attached to the base or to a central column of the solitary 1-celled (rarely 3 to 5 celled) pod, which opens at the top by as many or twice as many teeth as there are styles. The numerous seeds often kidney-shaped, and embossed with tubercles. The embryo in most of our weeds is curved so that the apex and base come close together around a central mass of albumen. The seeds do not develop mucilage when soaked in water. There are two tribes of these plants: the Cockles, with the sepals united into a tube, and the Chickweeds, in which the sepals are distinct or nearly so.

Plate II

COW COCKLE

Vaccaria Vaccaria. (L.) Britton

PLATE 11.

COW COCKLE, *Vaccaria Vaccaria* (L.) Britton.

Other English names: Cowherb, China Cockle.

Other Latin names: *Saponaria Vaccaria*, L.; *Vaccaria vulgaris*, Host.

(Noxious: N.W.)

Annual. Introduced from Southern Europe. Stem simple, branching above or much branched from the base, 1 to 2½ feet. Whole plant, smooth, succulent and glaucous (gray-green like a cabbage). Leaves ovate-lanceolate, clasping the stem. Flowers pale rose-pink, ½ inch across, in loose corymbose cymes. Calyx ovate, 5-ribbed and much inflated, and wing-angled in fruit. The smooth roundish capsules contain about 20 round, hard, dull black seeds [Plate 55, fig. 48—natural size and enlarged 4 times] about $\frac{1}{12}$ of an inch in diameter. These seeds are often confounded with and spoken of as the seeds of wild vetches; they can be told from these of a similar size, by their minutely roughened surface, or by cutting them open after soaking, when the entirely different embryo which lies in a circle around the seed just beneath the seed-coat will be seen. It in no way resembles the inside of a vetch or pea, which have two easily separable seed leaves.

Time of Flowering: July; seeds ripe in August.

Propagation: By seeds only.

Occurrence: As a troublesome weed, only in the Prairie Provinces.

Injury: Abundant in grain fields in Manitoba, and the North-western Provinces. This succulent plant absorbs much moisture, and its branching head crowds out crops. Prof. Hitchcock quotes Prof. Aven Nelson, that in 1896 it was the worst weed of Wyoming grain fields, specially in spring wheat; and Prof. Crandall reported that in 1893 it was so abundant in grain fields of Colorado as to rank among the worst weeds. (Nevada Weeds III. Nevada Agr. Exp. Station, Bul. 38.)

Cow Cockle is particularly abundant and injurious in Southern Manitoba. The seeds are also suspected of having poisonous properties similar to those of Purple Cockle.

Remedy: As the seeds do not ripen very early, early summer fallowing at short regular intervals will control this weed. The plants when in bloom are very conspicuous and often large. Hand-pulling at that time has been found very effective in the Mennonite settlements. The thorough cleaning of all seed grain is ery advisable with this weed, as the indications are that the seeds do not retain their vitality very long. Seeding down would also crowd out any plants which came up the first year, and prevent them from forming many seeds. Harrowing growing grain crops kills many of the tender succulent seedlings.

PLATE 12.

NIGHT-FLOWERING CATCHFLY, *Silene noctiflora*, L.

Other English name: Sticky Cockle.

Other Latin name: *Melandrium noctiflorum*, Fries.

(Noxious: Dom.)

Annual and winter annual. Introduced. Erect, 1 to 3 feet high, somewhat branching; whole plant covered with soft spreading glandular hairs. Lower leaves obovate, narrowed at the base; stem leaves lanceolate. Flowers few, erect, in a branching cyme, nearly an inch across, pinkish inside, yellowish white outside, opening at night; petals deeply divided; styles 3; capsule elongate ovoid, with 6 apical teeth; calyx at first cylindrical, afterwards broadly ovoid, with 5 long teeth at the apex, and marked with 10 prominent green nerves. Seeds [Plate 53, fig. 10—natural size and enlarged 8 times] in size and ornamentation much as in Bladder Campion, grayish brown, with a tiny black tip to each tubercle.

Time of Flowering: June to autumn; seed ripening in July and till frost.

Propagation: By seeds only.

Occurrence: Throughout the Dominion. Very abundant in the East, particularly in thin clover fields and in gardens; rather rare in the West, where it is a comparatively recent introduction.

Injury: A rank grower and a heavy seeder. The greatest injury is perhaps due to the frequency of the seeds in clover and grass seed, and the great difficulty in separating them.

Remedy: This weed is easily destroyed by ordinary methods, and, like all weeds, yields to a regular short rotation of crops. In clover grown for seed it should be hand-pulled as soon as the flowers show. Every care should be taken to sow high class clover seed free of the seeds of this Cockle.

Plate 12

NIGHT-FLOWERING CATCHFLY OR STICKY COCKLE
(*Silene noctiflora* L.)

Plate 13

WHITE COCKLE
(Lychnis alba, *Mill.*)

PLATE 13.

WHITE COCKLE, *Lychnis alba,* Mill.

Other English names. Evening Lychnis, White Campion.

Other Latin names: *Lychnis vespertina,* Sib.; *L. dioica,* L.; *Silene pratensis,* Godr. and Gren.

(Noxious: Dom.)

Sparingly introduced in Ontario. Biennial or short-lived perennial. Rootstock thick, sending up a few short barren shoots and long decumbent branching flowering stems 1 to 2½ feet high. Whole plant rather viscid hairy, not so much so as in the Sticky Cockle, which plant it resembles somewhat, but is wider branching, has many stems, the leaves are larger and the flowers, which are much more numerous, are pure white, with a more conspicuous crown of short white scales around the centre; the male and female flowers are on separate plants. Styles 5, not 3 as in Sticky Cockle, and the capsule has 10 teeth at the top instead of 6 as in that species. When mature, the calyx containing the capsule is in the White Cockle much larger and more swollen; the seeds [Plate 53, fig. 11—natural size and enlarged 8 times] are paler gray and rather larger than in Bladder Campion and Sticky Cockle.

Time of Flowering: June; seeds ripe in July.

Propagation: By seeds only.

Occurrence: Grain crops and meadows. The White Cockle is by no means a common weed in Canada, but has been introduced occasionally with crop seeds imported from Europe. In the vicinity of Guelph, it is abundant and troublesome, and it also occurs in a few other places in western Ontario.

Injury: Crowding crops; seeds an impurity in grass and clover seeds. This is a very persistent weed. Professor Day, of the Guelph Agricultural College, states that the roots are fleshy and hard to kill, unless dragged right up to the surface of the soil. If there is a little earth covering any part of them, they will grow and try to produce seed. When spudded below the surface, they will grow again but do not produce seed that season; persistent spudding is effective.

Remedy: A regular short rotation with frequent introduction of hoed crops. In meadows mow often to prevent seed from forming.

PLATE 14.

PURPLE COCKLE, *Lychnis Githago,* Lam.

Other English name: Corn Cockle (used in England, where wheat is generally spoken of as "corn").

Other Latin name: *Agrostemma Githago,* L.

(Noxious: Dom., N.W.)

Introduced. Annual and winter annual. Erect, 1 to 3 feet high; branches few; whole plant covered with soft silky hairs; not viscid. Leaves long and narrow, pointed, 2 to 5 inches long. Flowers purple, at the tips of the stems and branches, 1½ inches across; the petals notched at the apex, paler toward the center; calyx ovoid, much swollen in fruit, with the ribs very prominent, and the teeth long and conspicuous. Capsule ovoid, with five teeth at apex. Seeds [Plate 55, fig. 49—natural size and enlarged 4 times] pitchy black, varying from $\frac{1}{12}$ to $\frac{1}{8}$ of an inch in diameter, somewhat flattened, rounded triangular; the thin edge notched by the scar of attachment. Rough, covered with rows of short teeth.

Time of Flowering: July; seed ripe in August.

Propagation: By seed.

Occurrence: Grain fields.

Injury: An impurity in grain. The seed when ground with grain discolours the flour and renders it unwholesome, owing to the poisonous principle sapotoxin, which is found in this plant and some other Cockles.

Remedy: Thorough cleaning of seed grain. Hand-pulling when in small quantity. In districts where fall wheat is sown extensively, spring grains should be substituted for some time.

Plate 14

PURPLE COCKLE
(Lychnis Githago, *Lam.*)

Plate 15

CHICKWEED
(Stellaria media, *Smith*)

PLATE 15.

COMMON CHICKWEED, *Stellaria media*, Smith.

Other English name: Chickweed.
Other Latin names: *Alsine media*, L.; *Stellaria media*, With.

Introduced. Annual. Succulent, diffusely branching, decumbent. Roots hair-like and exceedingly tough. Leaves ovate, the lower ones with ciliate-hairy petioles. Stems bearing a conspicuous stripe of articulated hairs down one side. Flowers ¼ inch in diameter, star-shaped, numerous, solitary from the axils of the leaves, in old plants in terminal leafy cymes; petals white, about the length of the thin-margined calyx lobes. Capsules conic-ovoid, spreading or deflexed, longer than the calyx. Seeds [Plate 53, fig. 12—natural size and enlarged 8 times] small, $\frac{1}{24}$ of an inch in diameter; wedge-kidney-shaped, flattened and coarsely tuberculate, the tubercles arranged in regular curved rows, about 5 on each side and 4 on the edge. Colour, yellowish brown to dark brown.

Time of Flowering: At all times of the year, except during frost; seed ripening continuously.

Propagation. By seed.

Occurrence: This well known little weed occurs in all parts of Canada where the soil is moist and rich.

Injury: Choking out smaller and weaker plants, including seedlings of all crops. A persistent grower requiring constant hoeing to keep it down. Seeds often found in grass and clover seeds.

Remedy: Constant hoeing.

An occasional weed in the Maritime Provinces is the Grass-leaved or Lesser Stitchwort, *Stellaria graminea*, L., a wide branching plant 1 to 2 feet high, with many grassy leaves in pairs along the slender stems and bearing many starry white flowers nearly ½ inch across. The seeds [Plate 53, fig. 13—twice nat. size and enlarged 8 times] are often found in clover and grass seeds; they are of the same size but more nearly circular in outline and rounder in contour than those of the Common Chickweed; the surface markings are quite different; instead of tubercles, the surface is thickly covered with short curved ridges in more or less regular rows.

Somewhat similar to the Common Chickweed but easily distinguished from it, are the Mouse-ear Chickweeds, two or three of which occur in Canada as agricultural weeds. These are plants of much the same habit as the above, but covered all over with downy hairs, which in some species are glandular, giving a dirty appearance to the plants by reason of the dust which adheres to them.

The Common Mouse-ear Chickweed, *Cerastium vulgatum*, L., is a perennial plant which occurs in cultivated land, pastures and lawns throughout Canada. The pods are much elongated and curved upwards. The seeds [Plate 53, fig. 14—natural size and enlarged 8 times] resemble those of the Common Chickweed but have the tubercles fewer, less regularly arranged and more of the nature of short ridges than of low raised prominences as in the former species. The shape too is rather more angular, and they are not much more than half the size.

The FIELD CHICKWEED, *Cerastium arvense*, L., is in some places a troublesome and persistent weed. A native form occurs abundantly throughout the western prairies, but gives little trouble. In some parts of Ontario, Quebec and the Maritime Provinces there is a form with smoother leaves, which produces a copious system of underground rootstocks, which enables this plant to become a persistent enemy. Pastures or meadows invaded by it must be broken up and cleaned by a short rotation. The flowers of the Field Chickweed are large and conspicuous, more than ½ inch across, and borne on erect flowering stems 3 to 6 inches high. In the West the plant is sometimes grown as a garden flower for its beauty. The seed is larger than those of the preceding Chickweeds, almost round and coarsely tuberculate, with rounded prominences.

Plate 16

BLADDER CAMPION
(Silene inflata, *Smith.*)

PLATE 16.

BLADDER CAMPION, *Silene inflata,* Smith.

Other English names: Cow-bell, White Bottle.

Other Latin names: *Cucubalus Behen,* L.; *Silene Cucubalus,* Wibel.; *Silene vulgaris* (Moench) Garcke; *Behen vulgaris,* Moench.

Introduced. Perennial, with deep running rootstocks, which send up many barren shoots, and decumbent branched flowering stems. Whole plant pale green and in the common form perfectly smooth. Stems 1 foot to 18 inches high, forming large tufts. Leaves ovate-lanceolate, in pairs, meeting round the stems. Flowers white, nearly an inch across, drooping, the petals deeply divided. Calyx much inflated, pale green, veined with light purple, 5-toothed at the contracted apex. Capsule globular-ovoid, included in the calyx, opening by 5 short recurved teeth. Seeds [Plate 53, fig. 9—twice natural size and enlarged 8 times] round-kidney-shaped, about $\frac{1}{20}$ of an inch across, covered with concentric rows of small conical tubercles. The seeds of this species and of *Silene noctiflora* and *Lychnis alba* are so similar that they can be separated only by an expert. In many instances some seeds of one species resemble those of one of the other two so much that they are indistinguishable. In Plate 53, seeds of the three species have been represented which seemed best to show the average characters of each.

Time of Flowering: May to July; seeds ripe in July.

Propagation: By seeds and running rootstocks.

Occurrence: By roadsides, on railway banks and in hay fields, all through the Eastern Provinces.

Injury: The seeds are often found in clover and timothy seed.

Remedy: This deep-rooted perennial is difficult to eradicate. Deep ploughing and a short rotation of crops are necessary. Frequent cultivation with a broad-shared cultivator will be found very useful in holding this pernicious weed in check.

The young shoots of this plant have been used as a pot-herb, and are said to be excellent, having the flavour of both asparagus and green peas.

PLATE 17.

SPURREY, *Spergula arvensis,* L.

Other English names: Corn Spurrey, Sandweed, Pickpurse.

Introduced. Annual. Stems ascending, branching from the base, 6 to 18 inches high, almost smooth, sparingly glandular hairy above. Leaves narrowly linear 1 to 2 inches long, apparently whorled at the joints of the stem, but really clustered at the joints in two opposite sets of 6 to 8 together, with scale-like stipules between them. Flowers white, opening in sunshine, $\frac{1}{4}$ inch across, in terminal forked cymes; peduncles deflexed in fruit. Seeds lens-shaped or round and compressed, with the margins extended into a narrow wing, dull black, the marginal wing pale. The surface of the seed is more or less covered with small pale-coloured elongated protuberances, like gland-tipped hairs. These are sometimes entirely wanting, when the plant is called variety *sativa*. Both the small protuberances and the wing are sometimes lost by friction when the seeds occur among other seeds. The seeds [Plate 53, fig. 8—natural size and enlarged 8 times] vary much in size, but average about $\frac{1}{16}$ of an inch in diameter. The embryo is cylindrical and spirally coiled within the seed.

Time of Flowering: July; seed ripe July-August.

Propagation: By seed.

Occurrence: Occasional in fields and waste places throughout Canada, but frequent in grain fields in the Eastern Provinces, and in parts of British Columbia. Sometimes sown as feed for sheep or as a binder of soil on sandy land.

Injury: Troublesome on sandy land in the Maritime Provinces and in British Columbia. The seeds are frequently found in grass and clover seeds.

Remedy: Short rotation of crops. Frequent hoeing early in the season.

Plate 17

SPURREY
(Spergula arvensis, L.)

Plate 18

PURSLANE OR PUSLEY
(Portulaca oleracea, L.)

PLATE 18.

PURSLANE, *Portulaca oleracea,* L.

Other English names: Pusley, Wild Portulaca.

Introduced. Annual, of tropical origin, now found in gardens in most parts of Canada. Seeds germinating rather late. A fleshy prostrate perfectly smooth plant, freely branching from a single central root, with reddish stems and dark green alternate, wedge-shaped leaves mainly clustered at the ends of the branches. Flowers sessile, solitary, about $\frac{1}{4}$ inch across, with a two-cleft calyx and 5 small yellow petals; stamens 7 to 12, style 4 to 6-cleft. Capsule membranous, many-seeded, the top coming off as the lid of a box. Seeds [Plate 53, fig. 15—natural size and enlarged 8 times] black, roughened but shining, about $\frac{1}{40}$ of an inch in diameter, narrowly kidney-shaped, much as in the Pink family and, like the seeds of most of the members of that family, having the embryo curved and running around the outside of the seed.

Time of Flowering: July till frost, and seeds ripening for the greater part of that time.

Propagation: By seeds. The fleshy leaves and stems give it such vitality that flowering plants hoed out and left on the ground will continue ripening seeds for weeks.

Occurrence: In rich land and particularly in gardens. Most abundant in the Eastern Provinces; but being constantly introduced into new localities with seeds.

Injury: Although Purslane is an annual without any running root, and does not appear until late in the season, it is perhaps more difficult to extirpate than almost any other weed found on rich soil.

Remedy: Constant shallow hoeing, particularly when the young plants first appear, is the only way to control this weed. If left until the plants have attained a large size and the flowers have formed, they must be raked up after hoeing and removed from the land.

THE PEA FAMILY, *LEGUMINOSÆ*

This large and important family of plants is well represented in Canada and contains many useful food plants, such as peas and beans and the clovers, but also some very poisonous species, as the Loco Weeds, *Oxytropis,* and the Golden Bean, *Thermopsis*, of the western plains, as well as a small number of farm weeds of secondary importance. All plants of the Pea family serve a useful purpose as collectors of nitrogen from the air, which they render available for plant food.

Every species can be recognized as belonging to this family by one or other of two characters, both of which are peculiar to it, namely, a butterfly-shaped corolla, such as we find on a large scale in the Sweet Pea of our gardens, and, for a fruit, a pod like that of the same plant or of the garden pea, technically termed a *legume*. By far the larger number of the plants have both characters combined.

Mention may be made of the following, which are sometimes detrimental in farm lands. The WILD TARE, *Vicia angustifolia,* Roth, is an introduced annual in the Eastern Provinces, which, on account of its early ripening, is difficult to get rid of; the leaves are compound, of 8 to 16 linear or lanceolate leaflets; the flowers purple, 1 or 2 in the upper axils of the leaves; the pods are black and linear, with the tip sharp and turned up, 2 inches long, 4 to 12-seeded, the seeds [Plate 55, fig. 50—natural size and enlarged 4 times] globose, ranging from $\frac{1}{8}$ to $\frac{1}{10}$ of an inch in diameter, velvety black or olive brown mottled with white and dotted with fine black points. The Wild Tare is somewhat like the cultivated Tare, *Vicia sativa*, L., so valuable for fodder and known to some farmers as Vetches or Fitches. This latter has much larger leaves and seeds, with brown pods, and, what is of great importance when sown, does not persist in the land. To eradicate this pest a short rotation in which clover is included will be found most useful. After harvest, cultivation will help by causing many seeds to germinate which will be killed by the winter.

The PURPLE TUFTED VETCH, *Vicia Cracca*, L., a persistent perennial, is rather difficult to get out of old meadows but produces a large crop of rich fodder, which is rather beneficial than otherwise in hay.

The seed-pods of a few members of the Pea family become burs in wool, as for instance the Spotted and Toothed Medicks, *Medicago maculata*, Willd., and *M. denticulata*, Willd., neither of which, however, has established itself firmly in Canada. The only plant of this family, giving trouble in this way, is the Wild Liquorice, *Glycyrrhiza lepidota*, Pursh, which occurs on the prairies. The Sweet Clovers, *Melilotus alba*, Lam., and *M. officinalis*, Willd., often complained of by farmers, are biennial wayside weeds, which, as each plant lives for 2 years only, are easily subdued by preventing them from seeding. The Rabbit's-foot Clover, *Trifolium arvense*, L., is a useless member of the family which is not common in Canada and is of little importance.

THE ROSE FAMILY, *ROSACEÆ*.

The Rose family is more remarkable for ornamental plants than for those which may be considered agricultural pests. The family is extensive, and there are gathered together within its limits plants with regular flowers, by which their relationship can at once be seen but which present the greatest diversity of characters in the fruit. In the Spiræas or Meadowsweets we find the seeds in small, many-seeded, papery pockets called *follicles*. In the true roses the lower part of the calyx is fleshy and urn-like, holding many hairy achenes. In the raspberry the fruit is composed of many fleshy berries surrounding a dome-like receptacle. In the strawberry it is the receptacle which becomes fleshy and bears the seeds (achenes) sunk in little pits on its surface. In the Cinquefoils both the receptacle and the seeds are dry, and the markings on the hard shell of the latter are of great service in separating some of the closely allied species. In the species of Avens (*Geum*) and Dryas the achenes are prolonged into the persistent styles, which are jointed or hooked above or are feathery.

Of the small number of weeds belonging to the Rose family, it is sad to relate that one of the most troublesome is the beautiful Prairie Rose, *Rosa pratincola*, Greene, (which includes also *Rosa acicularis*, var. *Bourgeauiana*, and *Rosa arkansana* of Canadian writers). In Southern Manitoba this dwarf, large-flowered rose is found very persistent in grain fields by reason of its deep-rooted perennial underground stems, which send up many flowering shoots from the axils of scales on the rootstocks. To destroy Roses, the land should be ploughed with a sharp plough in hot weather and then disked twice afterwards at intervals of a week or ten days. Mr. T. N. Willing recommends spring ploughing rather than stubble cropping for land which is infested with Wild Roses and similar shrubs. The irregular, angular seeds, with hard, whitish shells [Plate 55, fig. 51—natural size and enlarged 4 times] are often found in the screenings of western grain.

The Hardhack, *Spiræa tomentosa*, L., is a pretty dwarf shrub, 2 to 3 feet, with short petioled, ovate, thick, toothed leaves, which are smooth above but downy beneath. The pink flowers are in dense steeple-like panicles at the tips of the erect branches. This shrub invades mountain pastures in the Province of Quebec and is very hard to keep down. Where pastures cannot be ploughed, the tufts must be pulled out by the roots and the spots sown with a quick-growing grass such as Orchard Grass, or the patches may be mowed down with a heavy, sharp scythe and grass sown. If there are enough sheep in the pasture, they will keep down the new shoots.

Silverweed, *Potentilla Anserina*, L., is sometimes found in damp land. It is a perennial with slender jointed runners, which root and form new plants at each joint. The leaves, silvery hairy beneath, are pinnate, with from 3 to 10 large ovate, sharply-toothed leaflets on each side, with very small ones between them. The long-stalked golden yellow flowers, nearly an inch across, are followed by a cluster of dry, smooth achenes. Silverweed roots on the surface of the land like a strawberry and is best controlled by draining the land and ploughing down the plants.

The Upright Cinquefoil, *Potentilla norvegica*, L., is an erect, branching, hairy annual with small yellow flowers and small kidney-shaped achenes with curved branching grooves on the surface. It grows commonly in meadows, and the seeds are frequent among grass and clover seeds. This is the *Potentilla monspeliensis*, L., of late floras.

THE EVENING-PRIMROSE FAMILY, *ONAGRACEÆ*.

There are a few weeds belonging to this family and many showy flowering plants, as the beautiful Fuchsias, Clarkias and Evening-primroses. The structure of the flowers is botanically of much interest; but, as the plants of this order which are troublesome as weeds are easily recognized, there is no need to speak of them here. The more noticeable of the weedy plants are, a few kinds of Willowherbs (*Epilobium*) which are also known as "fireweeds." The commonest kinds occurring on cultivated land are the Great Willowherb or Fireweed, *Epilobium angustifolium*, L., and the Sticky Fireweed, *Epilobium adenocaulon*, Haussk. These are rather persistent on wet land, the former from its running perennial rootstocks and the latter chiefly from the great number of downy seeds it produces. Of the Evening-primroses, two species require mention, the White-stemmed Evening-primrose, and the Common Evening-primrose, *Œnothera biennis*, L., with large yellow flowers. This tall, coarse biennial occurs throughout the country and is easily recognized by its tall branching habit (4 ft. by 3 ft.), its soft, downy lanceolate leaves and its large, showy yellow flowers, which open in the evening. This weed makes only a rosette of leaves the first year. For this reason it is a weed only in crops sown in autumn or on stubble. In thin clover fields it sometimes occurs conspicuously and should be either spudded out or cut off below the crown in the first season, or the tall flowering plants should be cut off below the surface of the soil and pulled out before the seeds ripen. This will disturb the soil much less than if the large, wide, spreading roots are pulled out. On stubble land to be sown to grain, the rosette-like plants should be destroyed by fall or spring cultivation.

The seeds [Plate 53, fig. 17, natural size and enlarged 8 times] are a frequent impurity in clover seed. They are produced in large numbers in long tapering 4-celled capsules which are clustered all along the stems. The dark reddish brown seeds are about $\frac{1}{16}$ of an inch in length, much angled by compression in the seed pods and with a roughened surface. As the pods do not easily shed their seeds and the plants are at all times conspicuous, much contamination of clover seed may be prevented with a little care at harvest time.

PLATE 19.

WHITE EVENING-PRIMROSE, *Anogra pallida* (Lindley) Britton, var. *leptophylla*, Nutt.

Other English name: White-stemmed Evening-primrose.

Other Latin name: *Œnothera albicaulis*, Nutt. and Canadian authors.

Native. Perennial. Stems mostly simple, shining white, sparsely pubescent above, somewhat decumbent, about 3 feet high; the leaves from 1 to 4 inches long, narrow and waved, but usually entire in our north-western plant. Roots white and fleshy, wide-spreading and throwing up flowering stems at intervals, thus forming large patches. Flowers axillary, large and handsome, $1\frac{1}{2}$ inches across, waxy-white turning pinkish as they fade, open in day time, odour unpleasant. Tips of the calyx-segments at the end of the buds free as 4 little points. Capsules narrow and curved, four-angled, about one inch long with the seeds in single rows in the 4 cells. Seeds [Plate 53, fig. 16—natural size and enlarged 8 times] about $\frac{1}{20}$ of an inch long, normally spindle-shaped but angular and somewhat twisted by compression in the pod, smooth and mucilaginous when soaked, yellowish brown (under the microscope, minutely dotted with black and faintly striate lengthwise).

This plant has usually been referred to in Canadian works as *Œnothera albicaulis*, but all the plants I have been able to examine are either the above named variety, or possibly *Œ. Nuttallii* (Spach) Rydb. True *Œnothera* (Anogra) *albicaulis* has the calyx segments closely joined together at the tips of the buds and the seeds are quite different, being bright yellow, lemon-shaped and pitted all over the surface.

Time of Flowering: July-August; seeds ripe September.

Propagation: By seeds and extensive deep-running fleshy rootstocks, every part of which when broken will throw out shoots and form new plants.

Occurrence: Sandy land; Manitoba and westward to British Columbia.

Injury: This deep-rooted perennial is very persistent in sandy land.

Remedy: Summer-fallow with deep or shallow ploughing, according to the nature of the soil, after the growth has been made in summer. Cultivate in fall or in spring before seeding to a crop.

Plate 19

WHITE EVENING-PRIMROSE
(Anogra pallida, *(Lindl.) Britton.*)

THE PARSLEY FAMILY, *UMBELLIFERÆ.*

This large family contains many herbaceous plants of weedy appearance, seldom of much floral beauty, but important as food plants, both for their large succulent roots, as in the Carrot and Parsnip, or for their fleshy leaf stalks, as Celery. The seeds of many are aromatic and wholesome, as Caraway and Coriander. There are also many plants like the Cowbane and Hemlock, which contain virulent poisons. The leaves are mostly pinnatifid, repeatedly sub-divided, and the flowers are borne in compound umbels, a form of inflorescence in which all the secondary foot-stalks of the flowers start from the top of the peduncle or general foot-stalk, like the supports of the ribs of an umbrella. The ovary is 2-celled and the tube of the 5-lobed calyx covers it and is completely adherent to it. The corolla has 5 petals, often unequal in size. When ripe, the 2 cells of the fruit separate into 2 seed-like halves, having 5 main ribs running lengthwise, which in the different plants are modified into wings or into rows of bristles or prickles of great value in separating the species. Inside the corky coat of the seeds are several longitudinal cavities filled with resinous or oily matters, which give the characteristic odours or flavours to the fruit. The true seed is inside the fruit and is tasteless. There are very few Canadian farm weeds belonging to the Parsley family; undoubtedly the most important members are the poisonous Cowbanes. The Caraway, *Carum Carui*, L., has run wild in some parts of Eastern Canada and is often complained of; but it is a biennial, and, if the plants are mowed closely or fed off for two years, it disappears. The Carrot and the Parsnip in the same way have escaped from cultivation in some places, but are easily disposed of in land that can be ploughed. Meadow land infested with Wild Carrot should be broken up and re-seeded.

PLATE 20.

SPOTTED COWBANE, *Cicuta maculata*, L.

Other English names: Cowbane, Water Parsnip, Water Hemlock, Poison Parsnip, Musquash Root, Beaver Poison.

Other Latin name: *Cicuta virosa*, L., var. *maculata*, Coult & Rose.

Native. Perennial. Stems stout, erect, hollow and jointed, widely branching, three to six feet high, quite smooth, pale green, dotted and streaked with purple. Leaves compound, twice or three times divided, clasping by an expanded base, the lower on long petioles, the upper sessile. The leaflets lanceolate, deeply toothed. Flowers small, white, in compound umbels, one to four inches across; the rays of the many-flowered umbellets unequal, from one to two inches long. Fruit [Plate 55, fig. 52—natural size and enlarged 4 times] smooth, ovate, compressed laterally, $\frac{1}{12}$ of an inch long, separating into two boat-shaped ribbed seeds. When cut across, these seeds show four oil tubes between the ribs and two on the flat side. Root, a bundle of a few fleshy spindle-shaped tubers, like small parsnips, at the base of the stem.

Time of Flowering: July to August; seeds ripe August to September.

Propagation: Copiously by seeds and by offsets from the crown of the root at the base of the old stem.

Occurrence: In low land along waterways, probably right across the Dominion.

Injury: Roots intensely poisonous to stock, particularly cattle, which pull them out when grazing in spring and eat them freely. When first turned out, the animals find few green plants to eat, and in browsing over the wet lands where these Water Parsnips grow they find the new green shoots, and when eating these pull out the roots. This is easily done owing to there being few root fibres. The roots not only look like small parsnip roots, but like them have a strong aromatic odour, which seems to make them very attractive to stock. It is claimed that the flowering plants when cut in hay may be eaten by animals without any ill effects, but that the ripe plants bearing seeds are dangerous. The whole plant, however, contains some of the poisonous principle, although it is true that this is most abundant in the roots and the seeds. Consequently no hay containing the Spotted Cowbane, or other Water Parsnips (also called Water Hemlocks), should be fed.

This plant, and in the West probably two or three other allied species closely resembling it (the Oregon Water Hemlock, *Cicuta vagans*, Greene, the Purple-Stemmed Water Hemlock, *Cicuta Douglasii*, C. & R., and the Wyoming Water Hemlock, *Cicuta occidentalis*. Greene), are the cause of nearly all the deaths of cattle reported in spring; and, unfortunately, in cases where much of the plant has been eaten no remedies can be applied. The means generally adopted on the plains in mild cases, when these are discovered in time, is to administer two or three daily doses of lard or bacon grease; but it is seldom that anything can be done on account of the intense virulence and quick action of the poison. A piece of root of the Oregon Water Hemlock, about the size of a walnut, is stated by Prof. Hedrick to be enough to kill a cow in about fifteen minutes.

Remedy: From the nature of the localities where Water Parsnips grow, hand-pulling is the best treatment for this dangerous weed. This is easily

Plate 20

SPOTTED COWBANE OR WATER PARSNIP

(Cicuta maculata, L.)

done, particularly if the roots are first loosened with a spud or some other implement. The plants should be carefully piled up to dry and then burnt or otherwise destroyed. The poisonous principle called cicutoxin is of a resinous or oily nature and will contaminate water, if, as is sometimes done, the pulled up plants are thrown into sloughs where they may be trampled upon by stock. It is most advisable that stockmen should know the appearance of these plants so as to destroy them whenever seen or, at any rate, so as to keep their animals away from localities where they grow too abundantly to be pulled out by hand.

THE SUNFLOWER FAMILY, *COMPOSITÆ*.

This, the largest family of flowering plants, includes many thousands of species and is represented in all parts of the world. The characters of the family are well marked. The flowers of all are composite, that is composed of many florets or small flowers standing together on an expanded enlargement at the ends of the stalks and known as the *receptacle*. Individually, these flower-heads or collections of many florets have the appearance of simple flowers and are popularly so spoken of, as for instance, the flower of the Sunflower, a Daisy, or a Dandelion, while in reality each one is a large number of flowers joined together at the end of a common footstalk, and what appears to be a calyx is a cluster of bracts or small leaves. A striking character of this family is that the anthers are united at their edges into a vertical tube with the style inside it. The calyx of the florets, when present, is united with the one-celled ovary, and in fruit is modified into a ring of silky bristles, awns, teeth or scales, which is called the *pappus*. The true seed is enclosed in a hard dry shell, like a small nut, botanically called an *achene*. The florets of composite flowers are of two kinds, both of which may sometimes be seen in the same flower-head, as in the Common Sunflower. The marginal or ray-flowers are strap-shaped and the smaller disk-flowers are tubular. When the flower-head has ray-flowers, either throughout or round the edge, it is termed *radiate;* when there are no ray-flowers it is said to be *discoid*.

Our Canadian members of this large family are divided by Dr. Asa Gray in his Manual, which is still used as the text book in most of our schools, into two series according to the nature of the corolla. In the first series, the Tube-flowered Composites, the corolla is tubular in all the perfect flowers, and regularly 5-lobed, strap-shaped only in the marginal or ray-flowers which, when present (as is not always the case), have either only pistils or have neither pistils nor stamens. In the second series or Strap-flowered Composites, the corollas on all the florets of the head are strap-shaped and perfect, that is, contain both pistils and stamens. To this series, known as the Chicory family (*Cichoriaceæ*), the Chicory belongs and many other plants with similar flowers, including many well known weeds, such as the Dandelions, Hawkweeds, Sowthistles and Lettuces. By far the larger number of our weeds belong to the first series, the Tube-flowered Composites; and in such large numbers of plants as are grouped under that one series, it becomes necessary to subdivide them into tribes.

In the ASTER TRIBE there are a few weeds of the Gumweeds and Fleabanes which are worthy of mention; but it is seldom that any of our Canadian species of the true Asters become aggressive weeds. The GUMWEED *Grindelia squarrosa*, Dunal, is a bright golden-yellow flowered plant of the western plains. This seldom becomes troublesome in crops, although the seeds [Plate 55, fig. 53—natural size and enlarged 4 times] have been found among wheat screenings and have occasionally been sent in under the impression that they were the seeds of Canada Thistle. They are, however, larger, and are much flattened and more angular, grooved lengthwise, and duller in colour. The Gumweed is accredited with causing hay-fever in the

West. The flower buds just before opening have a large drop of liquid resin on them, and, as the plants frequently grow along trails, they are troublesome from soiling ladies' dresses.

Of the FLEABANES, the Horseweed, *Erigeron canadensis*, L., is the most abundant and widespread, its tall spire-like stems being seen on stubble left for summer fallowing, on neglected land and in waste places in all parts of the Dominion. The Daisy Fleabane, *Erigeron annuus*, Pers., with coarsely-toothed leaves and the Rough Daisy Fleabane, *Erigeron strigosus*, Muhl., which has entire leaves, are common clover-field weeds in all parts of Eastern Canada. The small seeds are carried in grass seeds.

Among the EVERLASTING FLOWERS some species of *Antennaria* and *Gnaphalium* injure pastures by crowding out the grasses. By breaking up the sod these can be destroyed.

A small natural group consisting in Canada of the Ragweeds, Marsh-elders and Cockleburs, has been separated recently from the Aster family and called the RAGWEED FAMILY (*Ambrosiaceæ*); but in this publication it is thought better to keep it in its old place near the Sunflowers.

Some kinds of WILD SUNFLOWERS are noticeable weeds in the Prairie Provinces. In Manitoba the Many-flowered Prairie Sunflower, *Helianthus Maximiliani*, Schrader, and the Black-headed Sunflower, *H. rigidus*, Desf., are the most abundant; but the Wild Artichoke, *Helianthus doronicoides*, Lam., is the most difficult to eradicate. For all of these, early summer fallowing is the best method. The seeds are often found among western grain. In the Many-flowered Prairie Sunflower, they are [Plate 55, fig. 60—natural size and enlarged 4 times] about $\frac{1}{6}$ of an inch long, variable in shape, but mostly narrowly oblong egg-shaped in outline, flattened and rather angular, grooved lengthwise, brown, cross-mottled with irregular zigzag white lines; the apical and basal scars are both conspicuous, the latter rather oblique and indented in the middle.

On the western plains are many species of WORMWOOD, *Artemisia*, which are spoken of collectively as Sage brushes. Two of these, the Pasture Sage, *Artemisia Ludoviciana*, Nutt., and the so called Sweet Sage or Lesser Pasture Sage, *Artemisia frigida*, Willd., are sometimes troublesome by infesting home pastures, where the grass has been eaten close, necessitating the breaking up of the sod. The best known of the Wormwoods is the FALSE-TANSY, *Artemisia biennis*, Willd., a biennial which occurs in all parts of Canada and although very easily eradicated is a very unsightly weed when growing, as is often the case, among grain crops on stubble. The remarkably small seeds, $\frac{1}{15}$ of an inch, are dark brown, egg-shaped, wrinkled lengthwise and with a conspicuous pale-coloured ring-like basal scar. [Plate 53, fig. 18—twice natural size and enlarged 8 times.]

Closely allied with the Thistles is a weed which is rather abundant in meadows in the Maritime Provinces. This is the KNAPWEED, *Centaurea nigra*, L., a rather coarse perennial with thistle-like flowers, over one inch across and $1\frac{1}{2}$ inches high. The involucre or calyx-like whorl of bracts surrounding the flower-heads is spherical and composed of black-fringed scales. The seeds are about $\frac{1}{8}$ of an inch long by about half as wide, tapering to the base, cut off squarely above. On one side of the base, but above the end, is the large conspicuous basal scar; the large apical scar covers the whole of the top of the seed, and is surrounded by the pappus of two or three rows of short, flat bristles. The seeds (achenes) [Plate 56, fig. 63—natural size and enlarged 4 times] are slightly angular, somewhat flattened and striped lengthwise on their shining gray surface with pale ridges; the whole seed is sparsely hairy. The Knapweed is palatable to all stock and in no way injurious, but is unsightly and takes the place of the more valuable true grasses.

Plate 21

NARROW-LEAVED GOLDENROD
(*Solidago lanceolata, L.*)

PLATE 21.

NARROW-LEAVED GOLDENROD, *Solidago lanceolata,* L.

Other English names: Bushy or Fragrant Goldenrod, Yellow-weed.

Other Latin names: *Euthamia graminifolia* (L.) Nutt.; *Chrysocoma graminifolia,* L.

Native. Perennial. Stems erect, cymosely branched above, 2 to 3 feet high, almost smooth. Leaves numerous, linear-lanceolate, 1 to 5 inches long, the edges rough-pubescent. Separate heads of flowers about one-quarter of an inch across, bright golden yellow in dense, flat-topped clusters. Seeds (achenes) ovate-oblong, small, $\frac{1}{12}$ of an inch, downy. Pappus white.

Time of Flowering: July to September; seeds ripe in September.

Propagation: By seeds blown by the wind, and by long running rootstocks forming new plants at the tips, and, if left undisturbed, soon forming large patches.

Occurrence: In low land, throughout the Dominion.

Injury: Much complained of as a weed in damp hay meadows in the Eastern Provinces. The seeds, which are produced in large numbers, bear a silky pappus, by means of which they are blown long distances by the wind.

Remedy: This and all the other Goldenrods root near the surface of the ground and are easily destroyed by ordinary cultivation or shallow ploughing.

Several different Goldenrods are mentioned from time to time by farmers as rather troublesome, free-growing perennial weeds. The species most complained of is the Narrow-leaved Goldenrod; but frequent mention is also made of the Canada Goldenrod, *Solidago canadensis,* L., of which there are several varieties—the Smooth Goldenrod, *Solidago serotina,* Ait., and the Tall Hairy Goldenrod, *Solidago rugosa,* Nutt. These gay, showy autumn-flowering plants are all easily controlled by ordinary methods of good farming, and are more wayside and fence-corner weedy plants than agricultural pests of well worked land.

PLATE 22.

POVERTY WEED, *Iva axillaris*, Pursh.

Other English name: Small-flowered Marsh-elder.

Native. Perennial. Stems herbaceous, branching, ascending, from tough, woody extensive underground stems or rootstocks, 6 to 12 inches high, very leafy. Whole plant with a rank odour. Leaves thick, obovate to linear-oblong, entire, rough-hairy. The lower ones opposite, the upper alternate. Flower-heads drooping, solitary, on very short pedicels, from the axils of the upper leaves, $\frac{1}{6}$ of an inch across, inconspicuous. Seeds (achenes) pear-shaped, slightly flattened, sometimes keeled on the side and a little curved towards the base; colour variable, olive green, yellowish-brown to almost black; surface mealy and dull, $\frac{1}{8}$ of an inch long. Achenes [Plate 55, fig. 54—natural size and enlarged 4 times] very few, seldom more than one or two in each flower head, and very many heads have none.

Time of Flowering: June to August; seeds ripe July to September.

Propagation: Mainly by the extensive system of underground stems, which send up a great many flowering, leafy shoots.

Occurrence: In grain fields and pastures from Manitoba to the interior of British Columbia, thriving in all soils, but occurring generally on land where there is some alkali.

Injury: A most persistent perennial, forming large patches. Very exhaustive of moisture, thus starving crops and rendering the land hard to work.

Remedy: This has proved a most difficult enemy to dislodge when well established on the rich farms of the West. It requires, as Mr. T. N. Willing says in his Bulletin, "Hints to Grain Growers," Regina, 1905, well directed, persistent effort with sharp implements. The ploughing for summer fallow should be clean and deep, followed by frequent cultivation afterwards with a broad-shared cultivator. The seed of the low growing Poverty Weed seldom occurs in grain or grass seeds in our country, although Prof. Hillman found it in 11 per cent, of samples of alfalfa seed in Nevada.

The FALSE RAGWEED, *Iva xanthiifolia*, Nutt., a coarse annual with a remarkable superficial resemblance before flowering to the Great Ragweed, is a very common plant by roadsides, along railways and in corrals in Manitoba, where it grows to a height of 6 to 8 feet and produces an enormous quantity of seeds; these seeds [Plate 55, fig. 55—natural size and enlarged 4 times] are occasionally found among those of grain, grass and alfalfa from the West. They are of the same general shape as those of Poverty Weed, but are only $\frac{1}{10}$ of an inch long, more tapering and less robust, somewhat darker in colour, the surface finely striated lengthwise; when fresh, with a gray mealy covering which partially rubs off and gives them a mottled appearance. The young plant has the same habit of growth and leaf outline as the Great Ragweed, but can be recognized at once by taking hold of the stem, which in False Ragweed is perfectly smooth, while in the true Ragweed it and the leaves are noticeably rough; when full grown, the resemblance between the two plants disappears. The False Ragweed bears at the top of the stem a large, loose panicle of dark-coloured flowers, while the Great Ragweed has many of the leaves distinctly three-lobed and the tip of each branch ends with a long rat tail-like spike of male flowers.

Plate 22

POVERTY WEED
(Iva axillaris, *Pursh.*)

Plate 23

GREAT RAGWEED
(Ambrosia trifida, L.)

PLATE 23.

GREAT RAGWEED, *Ambrosia trifida*, L.

Other English names: Tall Ragweed, Crownweed, Kingweed, Bitterweed.

(Noxious: Dom., Ont., Man., N.W.).

Native. Annual. A tall, coarse branching plant, 4 to 8 feet high, with very rough stems and leaves; pale green and bearing the sterile and fertile flowers in different heads on the same plant, the sterile in long, slender spikes at the ends of the branches, and the fertile two or three together, sessile in the axils of the leaves at the base of the spikes. Sterile flowers cup-shaped, nodding, ¼ inch across; anthers yellow and conspicuous; fertile flowers inconspicuous; pistils slender and purplish. Leaves opposite on long margined petioles, very variable in shape, on young plants deeply indented but scarcely lobed; as the stems grow, 3 or even 5-lobed leaves are produced, and on many plants may be found leaves without lobes. Seeds (achenes) [Plate 55, fig. 57—natural size and enlarged 4 times] brown, urn-shaped, about ¼ inch long, tipped with a tapering beak and bearing around the base of this about one-third from the top, like the points of a crown, 6 or 8 blunt spines which are the ends of more or less distinct ribs; this crown-like appearance of the top of the seed has suggested the names Kingweed and Crownweed, sometimes used by millers.

Time of Flowering: July; seed ripe August.

Propagation: By seed, in grain and carried by water.

Occurrence: Ontario and occasional in other eastern provinces. Abundant in the rich Red River valley lands in Manitoba. Not extending west as a weed, but sometimes seen along the railways.

Injury: This coarse annual, when in crops, crowds and starves grain growing near it, but the chief loss to farmers is due to the difficulty experienced by millers in separating the seeds from grain, owing to its similarity in size and weight to wheat; the spines also are said to catch in the meshes of the screens and to give much trouble in the cleaning process.

Remedy: This is one of the few weeds in Manitoba for which hand pulling is a practical remedy. As a rule, the plants are conspicuous and grow near the edges of fields. A little labour in pulling before the seeds are ripe, being well repaid by the clean crop reaped, special attention should be given to fields liable to be flooded. Good work may frequently be done for this as for several other weeds by running a mowing machine around the edges of fields before the seeds are ripe.

PLATE 24.

COMMON RAGWEED, *Ambrosia artemisiæfolia*, L.

Other English names: Roman Wormwood, Smaller Ragweed, Hogweed.

(Noxious: Dom.).

Native. Annual. A coarse weedy branching plant, with hairy stems, 2 to 4 feet high. Leaves thin and much cut up, twice divided. Flowers and fruit [Plate 55, fig. 58—natural size and enlarged 4 times] much resembling those of the Great Ragweed but smaller. Occasionally plants may be found which bear only fertile flowers.

Time of Flowering: July; seed ripe August.

Propagation: By seeds, carried in the seeds of grain, clover and grasses.

Occurrence: In rich land and waste places throughout Eastern Canada and gradually extending into the Prairie Provinces.

Injury: The seeds are an impurity in clover, small grains and grass seed. The large spreading roots rob crops of moisture and plant food, and the free branching growth chokes out weaker plants.

Remedy: As the Ragweeds develop late in the season, roots and other crops should be hand-hoed after the usual horse cultivation. Land badly infested can be cleaned by a regular system of short rotations, care being taken to cultivate immediately after harvest, and to mow down the fall growth on new meadows.

THE PERENNIAL RAGWEED, *Ambrosia psilostachya*, DC., is a western plant found on the prairies, resembling the Common Ragweed in the shape of the leaves and flowers, but with running perennial rootstocks which throw up at intervals weak stems 1 to 2 feet high, covered with hoary-pubescent leaves. This Ragweed is seldom troublesome in Canada either as a weed on farm land or from the seeds occurring among crop seeds. With the more extensive cultivation of grasses and alfalfa for seed in the West, it may be expected that this plant may require more attention. The seed [Plate 55, fig. 59—natural size and enlarged 4 times] resembles that of the above very closely, but is, as a rule, more regularly oval and without the spines, although seeds bearing spines are not uncommon.

Plate 24

COMMON RAGWEED
(*Ambrosia artemisiæfolia*, *L.*)

Plate 25

STINKING MAYWEED
(Anthemis Cotula, L.)

PLATE 25.

STINKING MAYWEED, *Anthemis Cotula*, L.

Other English names: Mayweed, Dog's Chamomile, Dog-fennel.

Other Latin name: *Maruta Cotula*, DC.

Introduced. Annual and winter annual. Stems 12 to 18 inches, much branched from the root up, forming a flat topped bunch of white, yellow-eyed, daisy-like flowers, 1 inch across, on slender naked stems. Leaves twice divided, with the secondary leaflets cut into linear segments. Whole plant dull green, slightly hairy and with a strong unpleasant odour. Seeds [Plate 53, fig. 19—natural size and enlarged 8 times] dirty yellow, small, $\frac{1}{12}$ of an inch long, ovate-oblong or oblong, truncate at the upper end with a small knob in the center, abruptly pointed below, 10-ribbed with rows of coarse tubercles, sometimes however, according to Prof. Hillman, nearly smooth.

Time of Flowering: Summer to autumn; seed ripe by July, and young plants sometimes abundant in September.

Propagation: By seeds.

Occurrence: A common weed in old settlements, around buildings, along roads and in waste places, from the Atlantic coast to Manitoba, where it is as yet rare and only found along railways, but is rapidly appearing in new districts.

Injury: Frequent on cultivated land. The seed a common impurity in clover and grass seeds.

Remedy: Clean farming and the use of clean seed. Plants seen in clover fields should be pulled by hand.

In the Maritime Provinces the SCENTLESS MAYWEED, *Matricaria inodora*, L., is found commonly growing with the above. The two plants resemble each other closely, except that the Scentless Mayweed is a much handsomer plant with flowers nearly 2 inches across and foliage of a dark rich green. It lacks, however, the unpleasant odour and the seeds are entirely different. Although varying much in size and shape among themselves, these are similar in their structural markings. Prof. Hillman describes them, in his most valuable Bulletin, "Nevada Weeds," Part III, 1897, as "varying in length from one-sixteenth to one-twelfth of an inch; the smaller narrower ones are straight and rather prismatic in form, with truncate ends; large seeds are relatively broader and commonly curved in the direction of one face; they are sufficiently flattened to present two faces; three well defined broad ribs extend lengthwise along the usually concave face, broadly connected at the apex and meeting at the base; the opposite convex face exhibits the marginal ribs and a short partial central rib leading from the apex and connected with the marginal ridges below two deep depressions, which are separated by this rib. The surface between the ribs is black and transversely wrinkled." The apex of the seed is excavated and bears the scar of the flower at its center. The two depressions on the concave side give the seed a curious mask-like appearance. These depressions on the white unripe seed are seen as two green swollen translucent glands.

Yarrow or Milfoil, *Achillea Millefolium*, L., is a well known plant by waysides and in meadows everywhere from Atlantic to Pacific. In the West is a native form which occurs high up on mountains as well as on the prairies. The erect stems 6 to 18 inches high, bear flat-topped clusters of white flower-heads and finely divided fern-like leaves. The seeds [Plate 56, fig. 61—natural size and enlarged 4 times] are very small, flat and thin, gray with a white margin, about $\frac{1}{12}$ of an inch long and oblong-wedge-shaped; the basal scar is circular and distinct; the apical scar is a small cushion-like prominence from the middle of the broadly notched apex. These seeds are very often found among small grass seeds.

Plate 26

OX-EYE DAISY
(Chrysanthemum Leucanthemum, L.)

PLATE 26.

OX-EYE DAISY, *Chrysanthemum Leucanthemum*, L.

Other English names: White Daisy, White Weed.
Other Latin name: *Leucanthemum vulgare*, Lam.

(Noxious: Dom., Ont.).

Introduced. Perennial, shallow-rooted. Stems numerous, simple or little branched, 1 to 3 feet high. Basal leaves spatulate or oblong, crenate or coarsely toothed, narrowed into slender petioles; stem leaves sessile, partly clasping, deeply divided at the base and coarsely toothed above. Flower-heads solitary on long, naked peduncles, very handsome, $1\frac{1}{2}$ to 2 inches across; rays 20 to 30 pure white, spreading, 2 to 3-toothed at the apex; disk flowers yellow. Seeds club-shaped or elongate-ovate, $\frac{1}{12}$ of an inch in length, usually curved, almost straight on one side, and convex on the other, the knob-like scar at the top prominent; there are 10 well defined white ridges which run the whole length of the seed, meeting at both ends; between these ridges the surface of the seed is black minutely dotted with white; no pappus.

Time of Flowering: June; seeds ripe July.

Propagation: By short offsets from the woody rootstock; also more abundantly by seeds.

Occurrence: Enormously abundant in old pastures, in meadows and by roadsides from the Atlantic coast to the borders of Manitoba, and occasional along the railway to the Pacific coast.

Injury: A rank and aggressive weed in hay meadows, where it soon chokes out the grass. Seeds a common impurity in grass and clover seeds.

Remedy: Being a shallow-rooted perennial, the ploughing down of infested sod will kill all of the growing plants. A short rotation including seeding down to clover at short intervals, is probably the best method of cleaning land of this pernicious weed. The Ox-eye Daisy flowers at the time clover is ready to cut for hay, and, if this is done in good season, its seeds cannot ripen; when the sod is ploughed down, the old plants are destroyed.

A plant which is sometimes miscalled the "Ox-eye Daisy" is the beautiful BLACK-EYED CONE-FLOWER, *Rudbeckia hirta*, L., widely known as Yellow Daisy and Black-eyed Susan. This showy biennial weed is one of the exceptions to the usual direction of travel with introduced plants. Most of these have gone with civilization towards the West, but this denizen of the western plains has been brought eastward, probably with the seeds of grasses and clovers, and is now not uncommon in all provinces of Canada. It is a rather coarse rough-hairy biennial with long lanceolate undivided hairy leaves and with flower-heads of the same size as those of the Ox-eye Daisy, with glaring golden orange rays and a dark purple cone-shaped disk. The seeds [Plate 55, fig. 56—natural size and enlarged 4 times] are black, 4-angled, narrow, with parallel sides about $\frac{1}{8}$ inch long and without pappus. Meadows can be cleared of this weed by mowing, if this is done before the seeds are ripe.

PLATE 27.

LESSER BURDOCK, *Arctium Lappa,* L., var. *minus,* Gray.

Other English names: Bardane, Common Burdock, Clotbur.
Other Latin names: *Lappa minor,* DC.; *Arctium minus,* Schk.

(Noxious: Ont.)

Introduced. Biennial, from a deep thick tap root. Root leaves large, heart-shaped, downy beneath, somewhat resembling those of Rhubarb, petioles hollow. Flowering stem much branched, from 3 to 6 feet high, flowers purple, flower heads numerous, in clusters at the tips of the branches and in the axils of the upper leaves, $\frac{3}{4}$ inch across, oval-globular; the scales of the involucre ending with hooks by which the seed-bearing heads become burs and are distributed by becoming attached to passing animals, etc. Seeds [Plate 56, fig. 62—natural size and enlarged 4 times] oblong-ovoid, truncate at each end, flattened, with about 5 elevated longitudinal lines, generally somewhat curved, pale brown with dark transverse zig-zag depressed marks, apical scar circular with a central point, pappus, when present, consisting of several rows of short bristles barbed upwards.

Time of Flowering: July to August; seed ripe September.

Propagation: By seed.

Occurrence: Rich land in the older settled provinces; common in waste places, by roadsides and in orchards in sod.

There are two forms of the Burdock found wild in Canada as weeds; by far the commoner is the one figured in our plate, which has much more numerous and smaller flower heads, with the scales of the involucre shorter; the seeds are darker, sometimes showing hardly any pale brown marks; the petioles are hollow, which is not the case in the other and less common form, *Arctium Lappa,* L., (*Lappa major,* Gaertn., and L. *officinalis,* All., var. *major,* Gray). This latter has much larger green flower heads, $1\frac{1}{4}$ inches across, with the hooked scales more spreading and has longer peduncles.

Remedy: Cut below the crown or spud out when the ground is wet and soft, either the first year, or before the seeds are ripe in the second.

Plate 27

LESSER BURDOCK
(Arctium Lappa. *L.* var. minus, *Gray.*)

Plate 28

COMMON RAGWORT OR STINKING WILLIE
(Senecio Jacobæa, L.)

PLATE 28.

COMMON RAGWORT, *Senecio Jacobæa*, L.

Other English names: Stinking Willie, Baughlan, Tansy Ragwort, Staggerwort, St. James's-wort.

Introduced. Perennial, shallow-rooted and short-lived. Rootstocks few, short and thick from the base of the stem; under cultivation many plants after flowering the second year die without making any offsets. Stem stiff, erect, grooved, 2 to 3 feet; much branched above, forming a flat-topped dense compound corymb. Root leaves 6 to 8 inches long, petioled. Stem leaves sessile, embracing the stem; all leaves dark green, deeply twice-pinnatifid, the segments crowded and overlapping, crisped and waved. Flower heads numerous, erect and flat, golden yellow and very showy, ¾ inch across. Seeds [Plate 53, fig. 20—natural size and enlarged 8 times] creamy white, oblong, excavated at the top, with a small central point, deeply grooved along the sides; those of the disk with short bristles and almost straight, those of the ray-flowers smooth and much curved; pappus white. As is frequently the case in plants with composite flowers, this has seeds (achenes) of two different forms, according to their position in the flower head; those from the outside being more curved than those in the centre, which are straighter, narrower and more or less angled. Whole plant almost glabrous or with tufts of woolly hairs at the base of the leaves and flower heads, and with straggling hairs over the whole surface.

Time of Flowering: July till November; seeds ripe August.

Propagation: By seeds and by a few offsets from the base of the stem.

Occurrence: Locally abundant in Pictou and Antigonish Counties in Nova Scotia and in parts of Prince Edward Island, also reported from the Province of Quebec.

The history of this plant in Canada has been worked up by Dr. W. H. Pethick in Nova Scotia, and by Rev. Father Burke and Mr. L. W. Watson in Prince Edward Island; and good work has been done by all of these investigators in pointing out the danger of neglecting this weed. It would appear as if the Ragwort had been imported into the two Provinces independently, to Nova Scotia from Scotland, and to Prince Edward Island from Ireland, where it is known under the same name, Baughlan, as is used in its new home.

Injury: The chief injury by this weed is that, when eaten by cattle, it causes a curious and fatal disease of the liver (hepatic cirrhosis). For many years it had been suspected that the Ragwort was the direct cause of this malady; but recently this has been conclusively proved by a series of careful experiments carried out by Dr. W. H. Pethick, of Antigonish, under the direction of the Dominion Veterinary Director General, Dr. J. G. Rutherford, and the disease previously supposed to be contagious has now been removed from the list of affections dealt with under the Animal Contagious Diseases Act.

Remedy: The eradication of this coarse conspicuous and dangerous plant should certainly be possible on farm lands, now that land owners have proof of its true nature. It does not increase extensively from the root, and there is good evidence from the observations of Rev. Father Burke that where it has been mowed systematically it has in a short time disappeared. In pastures and hay meadows every plant should be grubbed out before the seeds form. Systematic and combined effort should also now be made by the municipal authorities to have it destroyed along roadsides and in the streets of towns and villages. In its green state, Ragwort is not readily eaten by domestic animals; therefore, the first attention should be given to cleaning meadows, because when mixed with hay it is eaten by all kinds of stock. Putting the land in the infested districts under a short rotation of crops, would at once do away with this short-lived and shallow-rooted perennial; and, with an effort to keep it mowed down in waste places, conspicuous benefit could not but follow at once. A short rotation of crops would also be a much needed improvement in the farming methods now practised in those parts of Nova Scotia and Prince Edward Island where the Ragwort is so abundant and injurious. It has been found that sheep can eat this weed with comparative impunity and that it dies out when closely fed off by these animals. It has been pointed out by Dr. Rutherford that the country where it occurs is admirably suited to the raising of sheep.

The COMMON GROUNDSEL, *Senecio vulgaris*, L., which has been introduced from Europe and is now found in gardens rather rarely in the Central and Prairie Provinces, but abundantly in the Maritime Provinces and on the Pacific Coast, is a small branching plant 6 to 8 inches high, bearing many tassel-like discoid or rayless flowers. The seeds are long and narrow, spindle-shaped, the upper end blunt and slightly enlarged by the white apical scar; surface finely ridged lengthwise, and covered with short white bristles, in this way differing from the seeds of the similar Stinking Groundsel, *Senecio viscosus*, L., which occurs in the Maritime Provinces with the Common Groundsel, and, although the whole plant is viscid pubescent, has its rather longer seeds entirely without bristles; its flower-heads also bear distinct marginal ray-florets.

The COMMON TANSY, *Tanacetum vulgare*, L., is quite a different plant from the Common Ragwort, although there seems to be much confusion in the Maritime Provinces as to the identity of the two plants. Tansy has almost rayless flowers, quite smooth 5-ribbed seeds with five blunt teeth at the top instead of a silky pappus. The plant, too, is pleasantly aromatic instead of rankly fetid, a character which in Nova Scotia has gained for the Common Ragwort the name of Stinking Willie.

Plate 29

CANADA THISTLE
(Cnicus arvensis, [illegible])

PLATE 29.

CANADA THISTLE, *Cnicus arvensis*, Hoffm.

Other English names: Creeping Thistle, Soft Field Thistle.
Other Latin names: *Carduus arvensis* (L.) Robs.; *Cirsium arvense*, Scop.

(Noxious: Dom., Ont., Man., N.W., B.C.)

Introduced. Perennial with very deep running rootstocks. Stems erect, 2 to 4 feet high, striate. Leaves very variable in shape, deeply pinnatifid, waved and crested, very prickly, in some plants much less so than in others, somewhat downy, particularly beneath the leaves. Flower heads very numerous, in a large loose corymb at the top of the stems, diœcious, some plants bearing male flowers only, which form no seeds, others female flowers only, which produce many seeds; the flower heads of male plants are nearly globular, 1 inch across, those of the female plants oblong, with short florets and only about half as large; large patches may be found bearing only male or female flowers, showing that each originated from a single seed. Flowers very variable in colour, ranging from pale purple through shades of pink to white. The seeds are ⅛ inch long, elongated oblong, smooth, somewhat flattened and curved, more or less bluntly angled and finely grooved; the top is squarely rounded and has a narrow rim with a small conical point in the centre, colour light brown; pappus copious, white. [Plate 56, fig. 64—natural size and enlarged 4 times.]

Time of Flowering: June to August; seed ripe July.

Propagation: By seeds and by extensive underground rootstocks, which send up both leafy barren shoots and flowering stems. This well-known pest, which, although called Canada Thistle in North America, is simply the Field Thistle or Creeping Thistle of England, has now been introduced into almost all of the British colonies, and has everywhere proved a troublesome and persistent enemy of farmers. As an instance of the rapidity with which this plant can spread, a single seed was planted on sandy but fertile land in the spring; by the first autumn there were two or three shoots showing at a distance of about a foot from the central plant; by the second autumn a patch 20 feet across with numerous stems had developed. The rootstocks naturally run down into the soil to a depth of 8 to 15 inches; but when patches are covered up stems can be thrown up through many feet of soil.

Occurrence: Abundant everywhere in Eastern Canada and Manitoba, here and there through the North-west Provinces and in British Columbia.

Injury: A gross feeder and a vigorous grower, which not only crowds crops but is a conspicuous evidence of negligent farming. Seeds frequent in grass and clover seeds, also in seed oats.

Remedy: Being a deep-rooted perennial, the Canada Thistle should be ploughed deeply in summer just as the flowers open, or the flowering stems may be mowed down and the land ploughed as soon as a new growth has appeared. As new stems are thrown up, they must be cut down with a broad toothed cultivator at intervals through the autumn. A deep ploughing late in autumn has been found very useful in killing Thistles and Perennial Sow-thistle in Manitoba. This would also be good practice in Eastern Canada, and the land might afterwards be thoroughly worked down and ridged up with a double mould-board plough for the winter.

The chief safeguard of farmers against Canada Thistle and all similar deep-rooted perennials is undoubtedly a regular short rotation of crops such as is described on page 11, as practised here at Ottawa by Mr. Grisdale. As an evidence of the above, it may be noted that the Canada Thistle is seldom seen in the crop on farms given to alternate husbandry, and it is everywhere more of a wayside weed which spreads on to farms when these are not properly worked.

A curious variety of the Canada Thistle in which the leaves are flat, almost entire and spinose only on the margins (*Cnicus arvensis*, Hoffm., var. *setosus*, Ledeb.), has been sent from Manitoba. This variety, however, has not proved stable, for plants grown from seeds of the variety can hardly be distinguished from the ordinary prickly form.

In addition to the Canada Thistle, two other species are sometimes complained of by farmers. The PRAIRIE THISTLE or WESTERN BULL THISTLE, *Cnicus undulatus*, Gray, is a native perennial with running rootstocks and large white-woolly leaves and stems; flower heads large, nearly 2 inches across, solitary at the end of the branches. This is not a very serious pest and soon disappears when the land is put under crop.

The BULL THISTLE or SPEAR THISTLE, *Cnicus lanceolatus*, Hoffm., is a large coarse biennial, introduced from Europe, 2 to 4 feet high, widely branching, with many large deep-purple flower heads 1½ inches high by as much across. This plant is very abundant in many parts of Canada, and particularly so in the partially cleared districts in British Columbia. As land is cleared and worked regularly, this weed, each plant of which lives only for two years, is killed out.

Plate 30

CHICORY
(Cichorium Intybus L.)

PLATE 30.

CHICORY, *Cichorium Intybus,* L.

Other English names: Wild Chicory, Wild Succory.

(Noxious: Dom.)

Introduced. Perennial from a long deep rootstock. Stems 2 to 3 feet high, branched, hairy below. Root leaves, closely resembling those of the Dandelion, 6 to 8 inches, spreading on the ground, runcinate pinnatifid or dentate; midribs hairy beneath; upper leaves glandular ciliate on the margins, clasping at base. Flower heads bright blue, nearly 2 inches across, composed entirely of strap-shaped flowers, usually closing by noon, in sessile clusters of three or four together along the almost leafless stems. Seeds [Plate 54, fig. 21—natural size and enlarged 8 times] $\frac{1}{8}$ inch long, dark brown or straw-coloured, mottled with brown, wedge-shaped, obtusely 3 to 5-angled, some seeds being much curved; the surface is grooved and ridged from top to bottom and roughened crosswise with minute close raised and waved lines; at the top, surrounding the apical scar, is a fringe of short flat white bristles.

Time of Flowering: July to frost; seeds ripe in August.

Propagation: By seeds or shoots from the roots.

Occurrence: Throughout Eastern Canada, most abundant in the Province of Quebec.

Injury: A troublesome weed in rich low land and in pastures. Seeds often found among crop seeds, particularly of clovers and grasses.

Remedy: A short rotation of crops, as for Canada Thistle. Chicory is not often seen in good farming districts except as a wayside weed.

PLATE 31.

FALL DANDELION, *Leontodon autumnalis*, L.

Other English names: August Flower, Autumnal Hawkbit.

Other Latin name: *Apargia autumnalis*, Hoffm.

Introduced. Perennial. Rootstock short and thick, frequently divided into several heads, each of which bears a thick tuft of runcinately toothed leaves somewhat resembling those of the Dandelion in outline, and several few-flowered leafless stems. Flower heads over 1 inch across, bright yellow. Seeds [Plate 56, fig. 65—natural size and enlarged 4 times] $\frac{3}{16}$ of an inch long, linear, a little swollen below, not beaked, ribbed lengthwise, the ribs connected by raised waved lines giving the whole surface a netted appearance; pappus dirty white, in one row of plumose bristles.

Time of Flowering: July till frost; seeds ripe August.

Propagation: By seeds and by division of the crown.

Occurrence: Abundant in the Maritime Provinces and parts of Quebec. Recently reported from hay fields in several places in Ontario.

Injury: Spreads rapidly from seed and overruns meadows, pastures and lawns, where it chokes out grass with its thick rosette of leaves in a very similar way to the COMMON DANDELION, *Taraxacum officinale*, Weber, the well-known pest of all long-settled districts. This latter, however, differs from the Fall Dandelion in having long deep tapering roots, every part of which, if broken off, will throw out leaves and form new plants, as well as in having hollow single-headed flower stalks and long-beaked green seeds. There is another Dandelion occurring with the preceding, the RED-SEEDED DANDELION, *Taraxacum erythrospermum*, Andrz., which differs from it merely by having reddish-purple seeds and more deeply divided leaves. The seeds of these Dandelions resemble in outline and size those of the Prickly Lettuce [Plate 56, fig. 67], but are not flattened as they are.

Remedy: The breaking up of pastures and meadows destroys all the plants of the shallow-rooted Fall Dandelion, which rot with the sod. In lawns the best treatment for all Dandelions is to dig out the plants and encourage the grass by sowing more seed, preferably of Kentucky Blue Grass and by light top dressings of quick-acting fertilizers, such as nitrate of soda, which may be used at the rate of 200 pounds to the acre, to be applied, not all at once, but in three or four light applications.

Plate 31

FALL DANDELION OR AUGUST FLOWER
(*Leontodon autumnalis, L.*)

Plate 32

ORANGE HAWKWEED OR PAINT BRUSH
(*Hieracium aurantiacum*, *L.*)

PLATE 32.

ORANGE HAWKWEED, *Hieracium aurantiacum,* L.

Other English names: Devil's Paint-brush, Paint-brush.

(Noxious: Dom.)

Introduced. Perennial. Low growing, throwing out many creeping branches close to the ground. Filled with bitter milky sap. Flowering stems 1 to 2 feet, erect and simple, almost leafless, bearing at the top a corymb of about a dozen handsome flower-heads nearly an inch across. The fiery orange-red of the flowers is very striking. Leaves spatulate or lanceolate, blunt-pointed, 3 to 8 inches long, tufted, many lying flat on the ground. Whole plant very hairy, the flowering stems clothed with stellate down, black gland-tipped hairs and long white hairs from black tubercles. Seeds [Plate 56, fig. 66—natural size and enlarged 4 times] small, $\frac{1}{16}$ to $\frac{1}{12}$ of an inch long, linear-oblong, cut off square at the top, pointed at the base, strongly 10-ribbed lengthwise, the tops of the ribs forming a star-like rim round the base of the dusky white pappus; colour of seeds purplish-black; unripe seeds deep red.

Time of Flowering: June; seeds ripe July.

Propagation: By seed and by creeping stems.

Occurrence: Abundant and very troublesome in the upland pastures of the Eastern Townships of the Province of Quebec and in some places in New Brunswick and Prince Edward Island. Reported occasionally from Ontario and all the Eastern Provinces.

Injury: A vigorous grower which spreads rapidly by means of its runners and matures a large quantity of small winged seeds, by means of which it soon overruns land that cannot be ploughed, the abundant and useless foliage taking the place of grass and ruining meadows and pastures.

Remedy: Although a vigorous grower, all the roots are close to the surface of the ground. In land used for crops, ploughing down and surface cultivation will kill it. Infested meadows and pastures must be broken up and put under a short rotation of crops. For mountain pastures or uplands where ploughing is difficult, the best treatment is that advised by Prof. L. R. Jones, of Burlington, Vermont, namely, to broadcast dry salt over the patches, so as to fall on the leaves of all the plants, at the rate of $1\frac{1}{2}$ tons to the acre (18 pounds to the square rod). This amount will kill all the plants of Hawkweed but will improve the grass.

BRANCHING HAWKWEED, *Hieracium cladanthum,* Arvet-Touve, MS. In hay meadows and pastures in parts of the Province of Quebec and in many places in New Brunswick, Nova Scotia and Prince Edward Island, there is a most pernicious and aggressive weed closely resembling the Orange Hawkweed, but having more numerous and rather smaller pale yellow flowers, narrower and longer leaves, and in strong plants tall flowering stems, sometimes three feet high, bearing a large irregular cymose panicle of flower heads. The lowest branches much elongated, given off even lower down than the middle of the stem, but raising up their flower clusters almost as high as those of the branches above them. On weak plants, such as grow in hay fields, this branching habit is much less conspicuous, and the form resembles much more nearly that of the Orange Hawkweed. The appropriate name of *Hieracium cladanthum*, or the Branching Hawkweed, has been proposed for this plant by the specialist in this genus, Mr. Arvet-Touve, of Gieres, France. It is just possible that this yellow-flowered Hawkweed, which is now widely known as the Yellow Devil, may merely be the yellow-flowered variety of the Orange Hawkweed, as the plants of the whole of this group are very variable, but it presents striking differences which seem worth characterizing.

Besides the above there is another species of the same group, the KING DEVIL, *Hieracium prœaltum*, Vill. (Noxious: Dom.), which is found with the above but much more rarely and which is far less aggressive with us in Canada, although it is bitterly complained of to the south of our borders; this is a much less leafy plant and bears fewer runners. The smaller flower heads are borne in the same way but the corymb is smaller and more spreading; the whole plant is less hairy, particularly the stems, and is glaucous. This grayish green colour makes the dark bases of the long hairs on the leaves stand out more conspicuously than on the allied species.

The MOUSE-EAR HAWKWEED, *Hieracium Pilosella*, L., has been introduced into Prince Edward Island since many years and has taken almost entire possession of some fields and extensive areas along the sides of the public roads. It is even a worse pest than the different species mentioned above. It is a mat-like prostrate plant, which produces long running leafy stems on the surface of the ground; these produce tufts of roots and side shoots at very short intervals, the latter being densely clothed with clusters of leaves, which are smooth above, except for some very conspicuous long white bristles, and are covered beneath with a thick felt of star-shaped hairs. The flower heads are solitary, on slender stalks, pale yellow, over an inch across and sweetly scented. The seeds are a little longer than those of the Orange Hawkweed, but the seeds of all the four Hawkweeds here mentioned are practically indistinguishable; all, however, are liable to occur in grass seeds of which they are a dangerous impurity. The agricultural treatment for the eradication of all is the same and consists mainly of a short rotation with seeding down to clover and grass at short intervals.

According to Dr. N. L. Britton (Flora of the Northern States and Canada, 1901), the Prince Edward Island Mouse-ear Hawkweed is the variety *Peleterianum*, Mer., of *Hieracium Pilosella*. The typical form of the species is found occasionally in the other provinces, having been introduced with European grass seeds.

Plate 33

BLUE LETTUCE
(Lactuca pulchella *D.C.*)

PLATE 33.

BLUE LETTUCE, *Lactuca pulchella,* DC.

Other English names: Showy Lettuce, Large-flowered Blue Lettuce.
Other Latin names: *Mulgedium pulchellum,* Nutt.; *Mulgedium acuminatum,* DC.; *Sonchus pulchellus,* Pursh.

Native. Perennial, deep-rooted. Stems 2 to 3 feet, leafy below. Whole plant smooth and glaucous, filled with milky juice. Leaves very variable, linear-lanceolate or oblong; entire, simply or runcinately dentate, or pinnatifid; stem leaves less divided and sessile. Flower heads nearly 1 inch across, pale blue, rather few, on scaly peduncles in a narrow panicle. Seeds [Plate 56, fig. 68—natural size and enlarged 4 times] about $\frac{3}{16}$ of an inch, one-quarter of which is a short thick beak; tip of beak expanded into a short cup-shaped disk; slaty gray when ripe, red when immature; club-shaped, flattened, with thick rib-like margins and with narrower ridges down each face; in some seeds one or more of these ridges much thickened, whole surface dull and rough; pappus pure white and silky.

Time of Flowering: June, July; seed ripe by end of July.

Propagation: By seeds and by fleshy running rootstocks.

Occurrence: Prairie Provinces and British Columbia. In moist or sandy soil, particularly where there is some alkali.

Injury: A deep-rooted and persistent perennial weed.

Remedy: Early summer-fallowing.

The PRICKLY LETTUCE, *Lactuca Scariola,* L. (Noxious: Man.) This coarse-growing prickly-leaved annual has spread rapidly through Canada during the past four or five years. Although, as a rule, it occurs most commonly in waste places, the seeds [Plate 56, fig. 67—natural size and enlarged 4 times] are frequently found among crop seeds. These are dark gray, similar to those of the black-seeded varieties of the garden lettuce, usually a little smaller, and like them are broadly lance-shaped and somewhat curved, flattened, margined and bearing 5 to 7 narrow ridges down each face; whole surface roughened, with very short white bristles on the ridges near the apex; beak as long as the seed, very slender and often twisted; pappus white. In the Eastern parts of Canada, where the Prickly Lettuce is now common in many places, plants average 3 to 5 feet in height; but in the Okanagan valley of British Columbia plants were found some years ago which were 8 feet high. The leaves are oblong-lanceolate, spiny margined and prickly on the midrib beneath, more or less pinnately divided, sessile with ear-like lobes at the base. The flower heads are pale yellow, less than ½ inch across and only a few open at a time on the large wide-spreading panicle.

The leaves of the stem are twisted at the clasping base so as to stand vertically with the edge to the sun, instead of horizontally, as in the case of the leaves of most plants. This peculiarity has given rise to a common name of this lettuce, the Compass Weed.

It is claimed by some botanists that the Prickly Lettuce, which is found in waste places in Canada, is the European *Lactuca virosa,* L.; but the two are difficult to separate with certainty. For the meantime it seems preferable to use the best-known name *L. Scariola.*

PLATE 34.

PERENNIAL SOWTHISTLE, *Sonchus arvensis*, L.

Other English names: Field Sowthistle, Creeping Sowthistle, Corn Sowthistle (in England), also sometimes incorrectly called the Russian Thistle, which however is an entirely different plant belonging to the Spinach family.

(Noxious: Dom., Man., N.W.)

Introduced. Perennial, very deep-rooted with large and vigorous running rootstocks. Stems 1 to 4 feet high, hollow, simple, with few leaves, and branching at the top. Whole plant filled with a bitter milky juice. Leaves 6 to 12 inches long, pointed, deeply cut, clasping the stem at their base and edged with soft spines. Flowers bright yellow, 1½ inches across, in corymbs, closing in strong sunlight; flower cup and flower stalk covered in the common form with long glandular hairs. A perfectly smooth and glaucous variety is common in parts of New Brunswick and also occurs at Port Hope, Ontario. Seeds [Plate 54, fig. 22—natural size and enlarged 8 times] brown, oblong, somewhat flattened, about ⅛ inch long, ridged both lengthwise and across, bearing at the top a copious tuft of very white silky hairs, which spread in drying and enable the seed to be carried long distances by the wind.

Time of Flowering: June to August; seed ripe July to September.

Propagation: Very rapid, by seeds and running rootstocks.

Occurrence: Abundant in cultivated fields and along roadsides from the Atlantic coast to Manitoba, where it is becoming very noticeable and giving much anxiety in the Red River Valley.

Injury: The Perennial Sowthistle, from its vigorous running rootstocks and the large amount of seeds it matures, is one of the most aggressive enemies of the Canadian farmer. Wherever it establishes itself, it causes great loss both in reducing the yields of crops and on account of the great difficulty in eradicating it.

Remedy: The adoption of a regular three-year or other short rotation of crops is the only hope of clearing a farm infested with this weed. This plan has been used with great success at Ottawa by Mr. Grisdale in cleaning a field badly infested with Perennial Sowthistle. In the West summer-fallowing must be done early and the land cultivated as often as necessary to keep down the fresh growths as they appear. If still abundant on the land, this should be plowed late in fall and seeded down the following year or used for a crop of oats or barley, to be sown late after spring cultivation and cut green for feed. When this plant first appears in new localities, the flowering stems should be pulled by hand as soon as the blossoms show in a growing crop so as to prevent the seed ripening. For land not under spring crop, plough deeply in June and follow with rape, late roots or buckwheat. Wherever Perennial Sowthistle is common a short rotation of crops is a necessity.

Plate 34

PERENNIAL SOW-THISTLE

(*Sonchus arvensis*, L.)

In addition to the Perennial Sowthistle, there are two smaller annual species of Sowthistles found commonly on rich land in all parts of Canada: the COMMON SOWTHISTLE, *Sonchus oleraceus*, L., and the SPINY SOWTHISTLE, *Sonchus asper*, L. The flowers of both of these are smaller, about ½ inch across, pale yellow. The two plants may be easily distinguished by the leaves and by the seeds. In the Common Sowthistle the leaves are more deeply cut and are arrow-shaped at the base, having spreading acute auricles or ear-like projections. The seeds [Plate 54, fig. 23—natural size and enlarged 8 times] are longitudinally ribbed and have fine cross lines connecting the ribs. In the Spiny Sowthistle the whole plant is more prickly, and the auricles at the base of the leaves, instead of spreading, are appressed to the stem and curve round beneath the leaf. The seeds [Plate 54, fig. 24—natural size and enlarged 8 times] are longitudinally ribbed but have no connecting cross lines. The two annual Sowthistles are particularly garden and stubble-land weeds. Land ploughed or cultivated after harvest is seldom infested. In gardens hoeing and hand-pulling before the seeds ripen will keep down these annuals.

THE BORAGE FAMILY, *BORRAGINACEÆ*.

The plants of this family are chiefly rough or bristly-hairy herbs with erect branching stems, simple entire leaves and symmetrical tubular flowers with 5 lobes and a 5-parted calyx. The ovary is solitary and when ripe divides in most species, into four hard one-seeded nutlets, which contain the true seed and which may be rough or highly polished, downy or armed with barbed prickles. The basal scars on the nutlets give useful characters in identifying them when found in commercial seeds. The flowers are usually in one-sided racemes that while young are coiled spirally inwards at the tips, but unroll as the flowers open and become nearly straight. The Borage family contains several weeds, as the Hound's-tongue or Common Bur, *Cynoglossum officinale*, L., a biennial wayside weed, the reddish purple flowers and barbed nutlets of which are familiar to everyone in Eastern Canada. The burs are very troublesome from getting entangled in the wool of sheep.

PLATE 35.

BLUEWEED, *Echium vulgare*, L.

Other English names: Viper's Bugloss, Blue-thistle, Blue Devil.

(Noxious: Dom.)

Introduced. Biennial, with a deep black tap root. Whole plant bristly hairy, the stiff bristles on the leaves from small pale prominences or tubercles, those on the stem from red ones. Flowering stems erect and wand-like, forming compound spikes of reddish buds and bright blue flowers, 1 to 2 feet high; the spikelets curved round at the tips as usual in the Borage family. Root leaves linear-oblong or linear-lanceolate, narrowed at base, entire, bristly hairy above and below, 6 to 8 inches long, forming the first year dense rosettes of long leaves lying flat on the ground; leaves of the flowering stems sessile. Flowers tubular-funnelform, rather irregular, with five rounded spreading lobes; calyx of 5 narrow bristly segments. Seeds (nutlets) [Plate 56, fig. 69—natural size and enlarged 4 times] 4 from each flower, dark brown, $\frac{1}{8}$ inch, irregularly angular-conical, hard and rough, sharply angled on the inner face and rounded on the outer with a keel running from the sharp apex half way down the outer convex face; basal scar triangular, large and flat, acutely margined, with a small deep hole close to the inner angle, having a little cone at the bottom; there are also two little conical projections beyond the centre in line with the side angles.

Time of Flowering: July to September; seed ripe August.

Propagation: By seed. Spread by dead plants being blown by wind in winter.

Occurrence: Common by roadsides and in waste places throughout Ontario and the eastern provinces. Chiefly on limestone and on gravelly or poor soil.

Injury: Most troublesome in rocky pastures. Seed occasionally found in clover and other crop seeds and as a bur in wool.

Remedy: This biennal weed is easily destroyed on cultivated land and is seldom seen in fields kept under crop, but is much dreaded and enquired about by farmers, presumably on account of its showy and weedy appearance.

Pigeon Weed, *Lithospermum arvense*, L. A seed which in form very much resembles that of Blueweed, and which is much oftener found in commercial seeds, is that of the Pigeon Weed, also called Red-root, Wheat-thief and Field Gromwell.

The size and general shape of the two seeds is similar. That of Pigeon Weed [Plate 56, fig. 70—natural size and enlarged 4 times] is rather less angular and the surface is much smoother. Instead of having rough projections, it is deeply and irregularly grooved, with prominent ridges between the grooves. It may be easily recognized by the basal scar, which is oval or

Plate 35

BLUEWEED OR VIPER'S BUGLOSS
(Echium vulgare, L.)

obtusely 3 to 5-angled in outline, and instead of the pit at inner angle and the two conical projections on an excavated surface, has a convex surface with two light-coloured distinct cones, one at the inner angle and the other opposite beyond the middle of the area. This weed is abundant in western Ontario and is troublesome in fields of fall wheat. It is a widely branching biennial or winter annual with white flowers, and produces a large crop of seeds, which ripen early. The seed is commonly found in poorly cleaned fall wheat and rye and also in timothy, mammoth clover and alsike seeds. Land is best cleaned by adopting a short rotation, including grain sown in spring instead of in autumn. Plants which germinate in autumn can be destroyed by fall or spring ploughing.

PLATE 36.

BLUE BUR, *Echinospermum Lappula,* Lehm.

Other English names: Stickweed, Sheep Bur.
Other Latin names: *Myosotis Lappula,* L.; *Lappula Lappula* (L.) Karst.

(Noxious: N. W.)

Introduced. Annual and winter annual. Erect, branching only above or from the base. Whole plant covered with short white hairs, which give it a grayish appearance. Leaves linear-oblong; root leaves about 3 inches, narrowed at base; stem leaves, sessile. Flowers small, pale blue, about $\frac{1}{8}$ inch across, erect, in leafy-bracted, more or less one-sided racemes. Seeds (nutlets) [Plate 56, fig. 71—natural size and enlarged 4 times] about $\frac{1}{8}$ inch, dark brown, pear-shaped, surface very rough; inner face sharply angled, outer face rounded, bare of spines in the centre but having on the sides a double series of long stiff spines, each of which has at its apex a star of from 3 to 4 sharp hooks.

Time of Flowering: From June; seeds ripe July.

Propagation: By seed only.

Occurrence: By roadsides and in waste places in the East. In the West chiefly in corrals and around buildings, but recently spreading with alarming rapidity into cultivated land, where it is sometimes abundant on fields left for summer-fallowing.

Injury: Seed very troublesome as a bur in wool; also frequently found as an impurity in commercial seeds, when many or all of the long barbed bristles may be rubbed off, but there is no trouble in recognizing it from the angled inner face with the small basal scar at the bottom of the central ridge, and the unarmed area on the outer face.

Remedy: Early summer-fallowing; fall or spring ploughing. Sow clean seed.

Plate 36

BLUE BUR OR STICKSEED

(Echinospermum Lappula. *Lehm*)

THE MORNING-GLORY FAMILY, *CONVOLVULACEÆ.*

The plants of this family possess marked characteristics by which they are easily recognized: *i. e.*, the twining stems, the trumpet-shaped flowers, and the globose cartilaginous capsules with distinct divisions. There are three species of Morning-glory, including among those plants which are classed as weeds, the notorious Field Bindweed, the Hedge or Great Bindweed, *Convolvulus sepium*, L., var. *americanus*, Sims, which sometimes gives trouble for a year or two after land is broken in the Prairie Provinces, and the Upright Bindweed, *Convolvulus spithamæus*, L., which occasionally is complained of in sandy land.

Closely allied to the Morning-glories are the Dodders, belonging to a small natural order, the *Cuscutaceæ*. These are curious leafless parasites occurring as loose tangled masses of fleshy threads, with clusters of flowers or small round pods at short intervals, attached to various plants, from which they draw their nourishment. The abundant occurrence of Dodder seeds among Alfalfa and Clover seeds offered for sale during recent years, as well as the presence of the parasite in large quantity in some fields of Alfalfa in Ontario, has given rise to much interest in these plants. The Dodder, the seeds of which are most frequently found as impurities with those of Alfalfa and Clovers, is the Alfalfa Dodder, *Cuscuta epithymum*, Murr., formerly known as *Cuscuta trifolii*, Bab., because it so often infested Clover (*Trifolium*). These seeds [Plate 54, fig. 34—twice natural size and enlarged 8 times] are small yellow or brown bodies varying in diameter from $\frac{1}{30}$ to $\frac{1}{20}$ of an inch, irregularly globose, and more or less angled on the inner scar-bearing side; the surface is granular-roughened, and these seeds are very apt to be overlooked from their lack of distinct characters. When soaked, the embryo is seen to be a spirally coiled worm-like body without any seed-leaves.

Mr. G. H. Clark, the Seed Commissioner of the Department of Agriculture, who has taken much trouble to investigate the origin of the Dodder seeds found in Canadian commercial seeds, has detected the seeds of another species in South American seeds, which he has identified as *Cuscuta racemosa*, Mart., var. *chiliana*, Eng. These seeds are about twice as large as those of Alfalfa Dodder and have a more rounded contour and a much larger and more distinct flat basal scar. These large seeds are, owing to their size, much more difficult to clean out of clover seeds than the common variety, and should be watched for carefully in clover and alfalfa seed.

PLATE 37.

FIELD BINDWEED, *Convolvulus arvensis*, L.

Other English names: Bindweed, Small Bindweed, European Bindweed, Small-flowered Morning-glory.

(Noxious: Dom., N. W.)

Introduced. Perennial, deep-rooting, with extensive creeping cord-like fleshy rootstocks; these throw up numerous slender branching and twining smooth stems, which form thick mats on the surface of the land, and twist around any plants growing within reach, using them as supports and choking them out. Leaves about 1 to 1½ inches long on slender stalks, ovate or heart-shaped, arrow-shaped at base. Flower-stalks slender, 1 to 2-flowered, about the same length as the leaves, bracted at some distance below the large open funnel-shaped pink flowers, which are over an inch across. Seeds [Plate 56, fig. 72—natural size and enlarged 4 times] rather large, $\frac{1}{8}$ of an inch, dark brown, pear-shaped; outer face convex, inner bluntly angled with flat sides. Surface roughened with small tubercles; basal scar, two smooth pale areas in a depression at the lower end of the inner face. Capsules globose, cartilaginous, 2-celled, containing 3 to 4 seeds. Embryo much folded and crumpled in the seed.

Time of Flowering: From June throughout the summer; seeds ripe in August.

Propagation: By seeds and by running rootstocks, every portion of which will produce new plants if broken up by the plough. It is noticeable that in many localities, as at Ottawa, this plant produces few seeds. It has, however, everywhere a most persistent habit of growth and deserves perhaps more than any other agricultural pest the appellation of "the worst weed in Canada."

Occurrence: Although very widespread throughout the Dominion and in restricted localities very troublesome, the Field Bindweed, fortunately, cannot as yet be called a common weed of Canada.

Injury: Exceedingly hard to eradicate, from its almost incredible persistence, owing to the vitality in the fleshy rootstocks.

Remedy: A short rotation including late sown roots or other hoed crops, rape being very useful for this purpose. Frequent use of a broad-shared cultivator will destroy new growths and exhaust the vitality of the plants. Great care should always be taken to sow no crop seeds containing those of the Field Bindweed. Applications of salt, lime, or straw, sometimes recommended to kill this weed are useless when used in practical quantities.

Plate 37

FIELD BINDWEED
(*Convolvulus arvensis, L.*)

THE FIGWORT FAMILY, *SCROPHULARIACEÆ.*

This large family, which includes the poisonous Foxglove (*Digitalis*) of Europe, does not contribute many farm weeds in Canada. The plants are remarkably variable in appearance, and in no order except perhaps the Orchids are there so many varieties of irregular corollas. The leaves present almost every form. The flowers are usually in spikes as in the Mulleins, or in panicles as in the rank-smelling Figworts, occasionally axillary and solitary as in the aromatic Musk. The Corolla is of 4 or 5 petals, either slightly united at the base or completely into a tube. The fruit is a 2-celled capsule which opens when ripe and allows the numerous seeds to drop out. In this family we find the Speedwells (*Veronica*), some species of which are troublesome weeds on lawns, the introduced Red Bartsia, *Bartsia Odontites*, Huds., and the Glandular Eyebright, *Euphrasia latifolia*, Pursh, both rather common pasture weeds in Prince Edward Island, and the persistent deep-rooted perennial weed the TOAD FLAX, *Linaria vulgaris*, Mill., common and injurious in all parts of Eastern Canada and gradually spreading into Manitoba. Where this plant has established itself, a short rotation of crops is essential. The showy pale yellow flowers with orange lips, nearly an inch long, are borne erect in dense racemes; the two-lobed corolla is closed and mouth-like, but, by gentle pressure at the sides, it opens and closes like the muzzle of an animal. The flat black disk-like winged seeds [Plate 54, fig. 30—natural size and enlarged 8 times] are about one-tenth of an inch in diameter and are often found in grass seeds. They are easily recognized from other seeds amongst which they occur, both by their shape and dark colour.

THE VERBENA FAMILY, *VERBENACEÆ.*

Is a small order allied to the Mint family and contains two or three tall herbaceous plants with small flowers, which are meadow and pasture weeds and of which the small seed-like nutlets are often found in grass seeds. The commonest of these is that of the BLUE VERVAIN, *Verbena hastata*, L. [Plate 54, fig. 31—natural size and enlarged 8 times]. It is brown in colour except the large whitish basal scar at the bottom of the inner face. The outer face is convex, irregularly ridged lengthwise and sharply angled at the sides. The inner face slopes to the margin from a sharply angled central ridge.

THE MINT FAMILY, *LABIATÆ.*

This large family contains several weeds, some of which are of common occurrence, but none of which are of much importance agriculturally. The characters of the family are well marked and easily recognized. A noticeable feature of these plants is the production of pleasantly aromatic and oily secretions as in the case of Lavender, Mint, Peppermint, Sage, Thyme, Rosemary, Bergamot, Patchouli, and many others which are used for the production of perfumes or in cooking. No plant in this large order is poisonous. Salient characters of this family are a square stem, often downy, a strong scent when bruised, leaves always opposite and simple, flowers generally clustered in the axils of the leaves, corolla irregular, more or less mouth-shaped with large wide open lips, the lower of which is generally much larger. The fruit of these plants consists of the 4 lobes of the ovary, which when ripe fall apart and become 4 seed-like nutlets or achenes lying loose at the bottom of the calyx. Few plants of the Mint family are troublesome enemies of the farmer on properly worked land, but the nutlets of a few are often found in commercial seeds. One of the most striking of these is that of the CATNEP, *Nepeta Cataria*, L. [Plate 54, fig. 33—natural size and enlarged 8 times], which is often found in Clover seed. It is about one-sixteenth of an inch long, reddish brown, roundly oval, a little flattened on the inner face. At the bottom of this face is the remarkable basal scar which has two clear white eye-like cavities, one on each side of it above.

The nutlets of the DRAGONHEAD, *Dracocephalum parviflorum*, Nutt. [Plate 54, fig. 32—natural size* and enlarged 8 times], and of the HAIRY MINT, *Stachys palustris*, Nutt. [Plate 56, fig. 73—natural size* and enlarged 4 times], have of recent years been found abundantly among the screenings of western wheat and occasionally also among the grain. These two seeds or nutlets are superficially much alike, but, when compared together, those of the Dragonhead are longer (⅛ inch) and narrower, being twice as long as wide, more angular and somewhat winged or wrinkled along the angles near the apex. The basal scar is large and curved, with a slit in the middle, giving it the appearance of a mouth; colour dark brown; the outer convex face very indistinctly ridged lengthwise and granular-roughened. In the Hairy Mint, the nutlets are smoother, dull black, rounded, nearly as wide as long, the inner face only slightly angled at the centre and edges, the scar merely a pale spongy spot at the sharp end of the nutlet.

Somewhat resembling at first sight the nutlets of Labiates are the seeds of the SUN SPURGE, *Euphorbia Helioscopia*, L. [Plate 56, fig. 77—natural size and enlarged 4 times]. When examined closely, however, they are easily recognized. They are rounded-oval in outline, rolling freely on a smooth surface, a little flattened on the inner face, with a central sharp ridge running to the apex; sides of the seeds not angled, as in those of the Mint family, scar kidney-shaped, white and very conspicuous, hollowed out at the base of the inner face; the whole surface of the seed coarsely netted with raised lines. These seeds have been found several times among vegetable and garden seeds.

THE RIBWORT FAMILY, *PLANTAGINACEÆ.*

This family in Canada embraces a few species of weedy stemless plants with inconspicuous flowers borne in long slender spikes at the top of bare flower-stalks or scapes. The fruit is a membranous capsule the upper part of which, when the seeds are ripe, drops off whole like the lid of a box. Although some species are very abundant on farm lands, they demand attention of farmers far more from the frequency with which the seeds are found among those of clovers and grasses than as troublesome weeds in fields. From their colour the seeds are very conspicuous among grass seeds, and on account of their size they are difficult to separate from them. This renders a knowledge of the appearance of the common kinds very important. The seeds of plantains may be roughly separated into two groups according to their shape: in one division, the seeds are irregularly angular like small grains of gun powder, *e.g.*, the Common Plantain and the Pale Plantain; in the other, they are boat-shaped, hollow on one side and rounded on the other, *e.g.*, the Narrow-leaved Plantain or Ribgrass, and the Bracted Plantain. All of these seeds develop a coat of mucilage when thoroughly wetted, by means of which their distribution is much facilitated.

* The figure to show the natural size is slightly too large.

Plate 38

COMMON PLANTAIN
(Plantago major L.)

PLATE 38.

COMMON PLANTAIN, *Plantago major,* L.

Other English names: Broad-leaf Plantain, Greater Plantain, Dooryard Plantain, Bird-seed Plantain.

Introduced and native. Perennial. Rootstock short and thick, erect, bearing many thick spreading roots and a large tuft of dark green oval, long-petioled, coarsely toothed, spreading or ascending leaves and several long dense spikes 3 to 12 inches long of inconspicuous flowers with purple anthers. Seed capsules oval, dividing about the middle. Seeds [Plate 54, fig. 25—natural size and enlarged 8 times] greenish brown, very variable in size and shape, according to the number in the capsule, which varies on different plants from 8 to 16, rounded on the outer face, angular on the inner or scar side; scar pale and conspicuous; the surface of the seed is finely netted with broken waved lines of dark brown, which radiate from the scar, average length one-twentieth of an inch.

Time of Flowering: May, throughout the summer; seed ripe in July.

Propagation: By seeds.

Occurrence: In various forms, some of which may be distinct species as indicated by the difference in habit and the degree of pubescence. Throughout the Dominion. Generally growing in rich moist soil.

Injury: Troublesome in meadows from reducing the grade of the seed and causing much extra expense in cleaning the seeds of grasses and clovers. On lawns the flat rosettes of leaves crowd out the grass and give an untidy appearance.

Remedy: Meadows to be left for seed should be thoroughly cleaned with hoed crops or by other special means and seeded down with well cleaned seed. In removing Plantains from lawns, a sharp knife should be run round deeply close to the crown, and the plant removed. This method disfigures the lawn far less than by digging out all the roots.

The Pale Plantain, *Plantago Rugellii,* Dec. Occurring with the Common Plantain, may often be found a rather larger plant with more erect smooth leaves of a paler or yellowish green, with the leaf-stalks purple at the base. The spikes are longer, and the flowers less crowded. The capsules more pointed, 4 to 9-seeded, opening below the middle. Seeds [Plate 54, fig. 26—natural size and enlarged 8 times] of the same angular shape as those of the Common Plantain, but about twice as large, and nearly black, with the surface merely roughened, not lined and netted. The seeds are a very common impurity in those of Timothy and Alsike.

PLATE 39.

RIBGRASS, *Plantago lanceolata*, L.

Other English names: Buckhorn, English Plantain, Ribwort.

(Noxious: Dom.)

Introduced. Perennial or biennial. Rootstock short and erect. Leaves numerous, 2 to 12 inches long, narrowly lanceolate and distinctly 3 to 5-ribbed, hairy and with tufts of brownish hairs at the base. In the first year the leaves lie close to the ground, forming a close rosette; on old plants they are erect. Flower stalks stiff, slender and grooved, 1 to 2 feet, much taller than the leaves. Flower heads at first ovoid and rather showy by reason of their numerous yellow anthers, elongating with age and forming dense cylindrical black spikes of seed, from 1 to 4 inches long. Capsules oblong, very obtuse, 2-seeded, opening about the middle. Seeds [Plate 54, fig. 28 —natural size and enlarged 8 times] chestnut brown, minutely granular-roughened, but highly polished, boat-shaped, with rounded ends, the outer face rounded with the edges folded inward around a deep longitudinal groove, in the centre of which lies the dark coloured scar which sometimes has a pale ring of dried mucilage around it.

Time of Flowering: Throughout the summer; seed ripe by July.

Propagation: By seeds.

Occurrence: From Atlantic to Pacific, but much commoner in some places than in others. Widely distributed with the seeds of clovers and grasses.

Injury: The chief injury by Ribgrass is due to the presence of the seeds among those of grasses and clovers grown for sale. The plant itself is palatable to stock and provides fodder of fair quality, although inferior to the true grasses.

Remedy: The ploughing down of infested meadows, and re-sowing with clean seed. The plants can be removed from lawns in the same way as the Common Plantain.

THE HOARY PLANTAIN, *Plantago media*, L. A plant which is much less frequently seen than Ribgrass but which has the same wide range, from the seed having been distributed with those of grasses, is the Hoary Plantain. This has the ovate leaves thickly covered with white hairs, short-stalked and always lying close to the ground in a dense rosette. It is deep-rooted and more difficult to eradicate from lawns by spudding than the other species here mentioned. The flower stalks are slender and about a foot high. Flower heads showy by reason of their purple filaments and white anthers, at first oval, gradually elongating to cylindrical spikes 1 to 3 inches long, flowers pleasantly fragrant. Capsules oblong, 2 to 3-seeded, seed of the boat-shaped class, of about the same size as that of Ribgrass, but thinner and flatter, often somewhat twisted, with the edges not so roundly turned in around the groove which bears the scar. Many seeds [Plate 54, fig. 27 —natural size and enlarged 8 times] show an indistinct shallow groove or

Plate 39

RIBGRASS
(Plantago lanceolata. L.)

constriction across the outer face just below the middle, indicating the part of the seed which fitted into the top of the capsule; these seeds have that part somewhat narrower and the margin more folded over the scar-groove.

The Hoary Plantain is intermediate in appearance between the Common Plantain and the Ribgrass, and its seeds are intermediate between those of Ribgrass and of the Bracted Plantain.

THE BRACTED PLANTAIN, *Plantago aristata*, Michx. This annual Plantain is a western plant which is rather uncommon in Canada, although its seeds are not uncommon in grass and clover seed in our seed trade. It has narrow linear grass-like leaves, the whole plant is downy, flower stalks erect, bearing thick cylindrical spikes 1 to 4 inches long, with conspicuous pointed bracts. Capsules 2-seeded. The seeds [Plate 54, fig. 29—natural size and enlarged 8 times] are boat-shaped, of the same size and form as those of Ribgrass, but are slightly wider, with sharper edges to the margin of the inner face. The elongated scar consists of two small shallow pits lying close together in the centre of the inner excavated face, the whole of which is whitened by a coating of dried mucilage. The rounded outer face has a distinct shallow groove crossing it just below the middle. This groove and the two pits of the scar present the best characters for distinguishing this seed.

THE SPINACH OR GOOSEFOOT FAMILY, *CHENOPODIACEÆ.*

The Spinach family contains many weedy plants, some of which are aggressive enemies of the agriculturist. Among these is the so-called Russian Thistle (Noxious: Man., N.W.), and also the Lamb's-quarters, which is probably the most abundant weed in all parts of the country. The flowers, which are almost always green and insignificant, have no corolla; each flower produces only a single seed within a bladder-like covering known as a *utricle;* but seeds are borne in enormous numbers on each plant. Some plants in this natural order supply wholesome articles of food, as the Spinach, Beetroot, Garden Orache and Lamb's-quarters.

PLATE 40.

LAMB'S-QUARTERS, *Chenopodium album*, L.

Other English names: Pigweed, Fat-hen, White Goosefoot.

Introduced and native. Annual. Extremely variable in every character. Mostly a tall succulent herbaceous annual with a slender, erect, grooved, much branched stem, 2 to 6 feet high, with angular-ovate, pale green, coarsely toothed leaves, narrowed at the base and borne on slender foot-stalks. Flowers in compound spikes from the axils of the leaves; whole plant more or less covered with white or pink mealy particles. Plants found late in the season are of a much darker green colour and have the leaves less angled. Seed [Plate 54, fig. 35—natural size and enlarged 8 times] about $\frac{1}{20}$ of an inch, circular in outline, more or less flattened on one side, strongly convex on the other; edges bluntly rounded; the lower convex face grooved from the margin to the central scar; seed shining black, minutely wrinkled, enclosed in a very thin papery seed vessel, called a utricle. The seeds, as found among crop seeds, have this thin seed vessel closely adhering to the seed as a brown or gray mealy deposit, which gives them a granular-roughened appearance; they also often have the dried 5-angled calyx closed tightly over them. When plants are picked or shaken roughly after the seeds are ripe, but while the plant is still green, the seeds fall out of the calyx very easily. Some seeds may also be found in screenings of grain, from which the brittle black coat has been partially broken away, when the yellow ring-like embryo will be seen surrounding the darker central portion of the seed.

A much larger seed, $\frac{1}{15}$ of an inch in diameter, of exactly the same appearance as the above, and which is sometimes found in crop seeds with it, is that of the MAPLE-LEAVED GOOSEFOOT, *Chenopodium hybridum*, L.

Time of Flowering and Seeding: From June to frost.

Propagation: By seed.

Occurrence: Everywhere, in rich land.

Injury: A gross feeder and a vigorous rapid grower, which in seasons favourable to its growth crowds and chokes out growing crops. Seed very abundant in all kinds of commercial seeds.

Remedy: Harrowing growing crops of cereals when the grain plants are three inches high will destroy myriads of the young seedlings of this and all other annual weeds among grain, which have germinated in the top 1 or 2 inches of soil, without injury to the much deeper-rooted drilled-in grain. When the plants are in small numbers or in clover grown for seed, pull by hand. Late plants growing in hoed crops should be carefully destroyed to prevent the seed from falling.

Plate 40

LAMB'S QUARTERS OR PIGWEED
(Chenopodium album, L.)

Plate 41

RUSSIAN PIGWEED
(*Axyris amarantoides*, L.)

PLATE 41.

RUSSIAN PIGWEED, *Axyris amarantoides*, L.

(Noxious: N.W.)

Introduced. Annual. A tall coarse plant from 2 to 4 feet high, erect and widely branching, very leafy. Stems grooved, rather pale at base, twigs and lower side of upper leaves covered with rusty pubescence. Leaves lanceolate, on short petioles, sparsely toothed. Flowers of two kinds; spikes of anther-bearing flowers, from half an inch to 3 inches long, terminate every branchlet, and fertile flowers cluster the branchlets thickly below these, each one producing a single seed. Seeds [Plate 56, fig. 74—natural size and enlarged 4 times] oval, flattened, $\frac{1}{10}$ of an inch, gray, with a silky lustre; surface minutely lined and wrinkled lengthwise; basal scar a short deep groove across the lower end; many seeds have the close-fitting utricle or cartilaginous covering projecting above the top as a two-lobed wing; this covering is beautifully mottled with white zig-zag lines on a brown ground. Embryo in a curved ring or loop around the outside of the central portion of the seed. Russian Pigweed when young has somewhat the appearance of Lamb's-quarters, but is a paler green, has a more wand-like habit of growth and, instead of being mealy, is softly pubescent with short star-shaped hairs. When full-grown, the whole plant forms a large pyramidal compound raceme; and, when mature, the stems, bracts and papery calyx segments turn white and make this weed very conspicuous.

Time of Flowering: June; seed ripe July-August.

Propagation: By seeds. Carried by the wind. The broken off plants and branches become tumbling weeds, by which the range of this Pigweed has been much extended.

Occurrence: The species was first noticed in Canada in 1886 by the roadside at Headingley, 10 miles west of Winnipeg in Manitoba, to which place it was said to have been brought direct from Russia. It is now found along the lines of railway throughout the North-west, and has even been detected on a railway bank as far east as Sherbrooke, P.Q.

Injury: A leafy, gross feeding, wide-rooting annual which crowds growing crops and gives a very weedy appearance to farm lands. The thick woody stems are very troublesome when crops are being harvested. The abundant seeds, somewhat like small gray flax seeds, are found commonly in grain from a few districts infested by this rank-growing weed. It is, therefore, most important that every care should be taken to prevent it from spreading from roadsides and waste places, as it has every characteristic of a bad and aggressive enemy.

Remedy: Harrowing out young seedlings from a growing grain crop, in the same way as Lamb's-quarters. Hand-pulling when not too abundant. Mowing and burning plants by waysides, along railways and in waste places.

THE PIGWEED FAMILY, *AMARANTACEÆ.*

The Pigweeds are a small family of plants, mostly of tropical origin, very closely allied with the Chenopods or Spinach family. All of them with us are coarse weedy annuals of rich land in gardens and on farms. The flowers are small and inconspicuous, and they produce enormous quantities of small highly polished, lens-shaped, more or less margined seeds. The leaves are simple and petioled. Some of the exotic members of this family are gorgeously coloured and are grown for their ornamental foliage or inflorescence, as the Cockscombs (*Celosia*), the Rainbow Amaranth, *Amarantus tricolor*, and the well known Love-lies-bleeding, *Amarantus caudatus*, L. The seeds are borne singly as in the Spinach family, and enclosed in a thin cartilaginous covering known botanically as a utricle.

PLATE 42.

REDROOT PIGWEED, *Amarantus retroflexus*, L.

Other English names: Rough Pigweed, Chinaman's Greens.

(Noxious: N.W.)

Introduced. Annual, with a rosy pink tap-root. Stems erect simple or branched, rough pubescent. Leaves long-petioled, ovate, bristle-pointed. Flowers inconspicuous, numerous, crowded on thick compound spikes at the ends of the branches and in the axils of the leaves. Bracts of the flowers bristle-pointed, longer than the green sepals. Seed [Plate 54, fig. 36—natural size and enlarged 8 times] highly polished, reddish black to jet black, about $\frac{1}{24}$ of an inch in diameter, circular or egg-shaped in outline, much flattened and equally convex on both sides; poorly filled seeds have a narrow slightly flattened marginal band, which marks the location of the ring-like embryo lying round the outside of the seed. The basal scar is a light point on the edge of the seed where the two tips of the embryo meet, and has a low depression around it on both faces of the seed.

Time of Flowering: July to September; seed ripe by August.

Propagation: By seed.

Occurrence: In all crops. Thoroughly established in all the settled portions of the Dominion. Abundant in waste places around farm buildings and in gardens. Widely spread by the seeds, which occur commonly in all commercial seeds.

Injury: A large weedy plant which crowds crops and increases the labour of working land. Seeds common in those of grasses and clovers.

Remedy: Can be controlled by shallow cultivation and hand-pulling.

The TUMBLE WEED, *Amarantus albus*, L. Another Amaranth which is very abundant throughout the country, but particularly so in the West, is the Tumble Weed, a bushy branched erect or procumbent annual weed with whitish stems and small oval or spatulate leaves, in the axils of most of which, on old plants, are small clusters of flowers or seeds. When mature,

Plate 42

REDROOT PIGWEED
(*Amarantus retroflexus*, *L.*)

these plants break off at the ground and are blown long distances by the wind, scattering their seeds as they go. The seeds are much like those of the Red root Pigweed, and although averaging rather smaller, about $\frac{1}{32}$ of an inch, cannot always be distinguished from them when found among other crop seeds.

SPREADING AMARANTH, *Amarantus blitoides*, Watson. This species resembles the Tumble Weed very much, but has rather larger rounder leaves and prostrate diffusely branching rather fleshy stems, which form large mats attached by the central root. This is a native annual plant of the western prairies, but is frequent along railways in the East, and the seeds are often found in alfalfa, clover and grass seeds from the Western States. The seeds can be easily distinguished from those of the Tumble Weed by their large size, $\frac{1}{16}$ of an inch, which is nearly twice that of the other species. Unless closely examined, the seeds of the Amaranths may sometimes be confused with rubbed seeds of the Lamb's-quarters. The surface of the former, however. is always more highly polished and smoother. The character of the scar will be found the easiest point for distinguishing them, being a central point with a long groove on one side in the Chenopods, and a notch in the margin in the Amaranths. The seeds of the Lamb's-quarters and of the Red-root Pigweed are small and of about the same size; those of the Maple-leaved Goosefoot and of the Spreading Amaranth are both nearly twice as big and compare with each other in a similar manner.

THE BUCKWHEAT FAMILY, *POLYGONACEÆ.*

The Buckwheat family contains several weedy plants, some of which are troublesome on farms. The objectionable members of the family all belong to two genera, the Docks (*Rumex*), and the Smartweeds, or Knotweeds (*Polygonum*). Many of these require some attention to keep them under control, but none can be said to be very difficult to eradicate on well-kept farms. The Docks are tall-stemmed perennial weeds with tap-roots, found in pastures and meadows; they live for many years but, with few exceptions, as in the case of the Sheep Sorrel, do not spread from the root. The achenes enclosing the seeds are shaped like those of the buckwheat, being triangular in cross section, or as the name buckwheat (which is merely a corruption of beechwheat) indicates, shaped like small beechnuts. There is no true corolla, the achenes being surrounded merely by the 6-parted calyx, 3 segments of which are small and the other 3 large, wing-like and variously shaped and veined in the different species. One or all three of these wings may bear a seed-like-corky tubercle on the outside. The "seeds" of all species are much alike, and the wing-like segments, often found attached to the seeds as they occur among crop seeds, are a great help in recognizing the different species.

The weeds of the Knotweed and Smartweed division of the family are found in three rather distinct groups of the genus *Polygonum*:

(1) The Knotweeds are well represented by the very common DOORWEED, *Polygonum aviculare*, L., which accompanies civilized man everywhere, and is found along roads and trails, forming mats of spreading wiry jointed stems with a leaf, a pair of silvery scales and a small cluster of flowers at each joint, each flower producing a slender reddish-brown triangular achene $\frac{1}{12}$ of an inch long.

(2) The Smartweeds, or Persicaries, are represented by several plants of various habits, some are perennials with extensive running rootstocks, as the SWAMP PERSICARY, *Polygonum Muhlenbergii*, Wat., which is sometimes very persistent in low undrained spots, others are fleshy-stemmed annuals, as the LADY'S-THUMB, *Polygonum Persicaria*, L., the shining black, $\frac{1}{12}$ of an inch, ovate heart-shaped, hollowed out on one side or roundly triangular seeds (achenes) of which [Plate 54, fig. 37—natural size and enlarged 8 times] are frequently found among the seeds of grasses and clover, and the DOCK-LEAVED PERSICARY, *Polygonum lapathifolium*, L., a common tall-growing and rather aggressive weed among grain and clover on rich low land in all parts of Eastern Canada. The "seeds" of the latter, which also occur with the above, are $\frac{1}{10}$ of an inch long, more roundly heart-shaped, chocolate brown, hollowed on both faces and never triangular. A larger ($\frac{1}{8}$ inch) blackish "seed" closely resembling both of the preceding, but more frequently bearing the spike-like base of the pistil at the tip, is that of the GLANDULAR PERSICARY, *Polygonum pennsylvanicum*, L. The "seeds" of this plant, as those of the Lady's-thumb, are sometimes triangular.

(3) The third group contains plants mostly with twining or climbing stems and with arrow-head-shaped leaves, as the Wild Buckwheat [Plate 43].

Plate 43

WILD BUCKWHEAT OR BLACK BINDWEED
(Polygonum Convolvulus. L.)

PLATE 43.

WILD BUCKWHEAT, *Polygonum Convolvulus*, L.

Other English name: Black Bindweed.

Introduced. Annual. A twining vine with rather rough branching stems and thin, smooth, arrow-head-shaped leaves. Flowers greenish, drooping, on short slender pedicels, in small clusters from the axils of the leaves and in loosely flowered terminal racemes. Calyx 5-parted, persistent, closely wrapped around the single dull black triangular seed (achene) [Plate 56, fig. 75—natural size and enlarged 4 times], which is about $\frac{1}{8}$-inch long, bluntly pointed at the apex, and almost twice as long as broad, widest just above the middle; embryo club-shaped, small, curved and lying along one angle of the seed in a groove in the large central mealy mass.

Time of Flowering: From June throughout the summer, the seeds ripening irregularly from about the beginning of July.

Propagation: By seed.

Occurrence: General. Most injurious in the Prairie Provinces.

Injury: Twining around the stems of the small grains, binding them together, pulling them down and choking them out, also a great nuisance in potato fields. The seeds begin to ripen long before all grain crops, and in that way land devoted to grain crops for several years becomes badly infested; the seeds are one of the most abundant impurities in grain sent to the market, particularly in wheat and oats. The seeds have considerable value as feed for stock, for which reason screenings containing these and other weed seeds are often carried back from the elevators by farmers and fed without grinding or scalding, which is a dangerous practice.

Remedy: Harrow or cultivate stubbles directly after harvest, so as to cover up and encourage the germination of as many seeds as possible in autumn. The young plants will be killed by frost. Those seedlings which germinate in spring must be destroyed by cultivating before seeding or by harrowing after the grain is up. Ploughing for summer-fallow must be done early, so as to turn down plants on stubble before the seeds ripen.

PLATE 44.

CURLED DOCK, *Rumex crispus,* L.

Other English names: Yellow Dock, Sour Dock.

(Noxious: Dom.)

Introduced. Perennial, with a very deep tap-root. Stem 2 to 3 feet, smooth, erect, terminating in wand-like racemes. Root-leaves oblong-lanceolate in outline with much crested or waved margins, 6 to 12 inches long, on long petioles; stem-leaves on short petioles and much smaller or absent towards the top of the stems. Flowers small, in rather widely separated clusters around the stems. The 3 inner segments of the calyx enlarging as the seed ripens, heart-shaped, with the margin entire or obscurely toothed, all with grain-like tubercles on the outside. Seed (achene) [Plate 56, fig. 76 natural size and enlarged 4 times], $\frac{1}{12}$ of an inch long, shaped like a miniature beech-nut, dark brown, shining.

Time of Flowering: June; seeds ripe July.

Propagation: By seeds. Clumps increasing slowly by shoots from the crowns of old plants.

Occurrence: In fields and waste places. Naturalized from Atlantic to Pacific, very abundant in Southern and Western Ontario.

Injury: A common weed in meadows and pastures, and also abundant by roadsides, whence the seeds blow on to cultivated land. The seeds are a common impurity in grass and clover seeds.

Remedy: Land worked under a short rotation is never badly infested by docks. In clover meadows all plants seen should be spudded out. This is easily done when the land is soft after rain.

Plate 44

CURLED DOCK
(*Rumex crispus*, *L.*)

Plate 45

SHEEP SORREL
(*Rumex acetosella, L.*)

PLATE 45.

SHEEP SORREL, *Rumex Acetosella*, L.

Other English names: Sour-grass, Field Sorrel, Red Sorrel.

Introduced. Perennial, very persistent by extensively spreading yellow fleshy rootstocks. Stems slender, 6 to 18 inches, erect or ascending, branched above. Leaves with silvery 2-parted stipules at the base, narrowly arrow-head-shaped, entire, 1 to 4 inches long, quite smooth and rather fleshy, on long petioles. Flowers numerous in panicled racemes, of two kinds on separate plants: the male or anther-bearing flowers have conspicuous anthers; the female or pistillate flowers are much less showy and are tipped with three tiny crimson feathery stigmas. Seeds (achenes) [Plate 54, fig. 38—natural size and enlarged 8 times], as they occur among clover and grass seeds, generally covered by the three larger conspicuously veined calyx segments, which fit closely over the seed. The three small segments which alternate with these, fit over the angles of the achene outside the edges of the larger segments. The achene itself when the calyx segments are removed is $\frac{1}{20}$ of an inch long and nearly as broad, triangular-ovate, pale brown, shining.

Time of Flowering: May to August; seeds ripe July to September.

Propagation: By seeds and shallow running rootstocks.

Occurrence: Naturalized in all parts of the country.

Injury: The seeds are one of the most abundant impurities in clover and grass seeds. The plants increase rapidly in thin or worn-out meadows and in pastures, both on uplands and in hay marshes—crowding out the grass and much reducing the crop. Sheep Sorrel is also a troublesome weed in gardens.

Remedy: In upland pastures a top-dressing of lime is said to be very beneficial. Where land can be ploughed the sod should be manured and ploughed down and the land re-seeded. In gardens, constant shallow cultivation is necessary.

THE GRASS FAMILY, *GRAMINEÆ.*

There are among the true Grasses not only most valuable fodder plants, besides all of the cereal grains, but also several bad weeds, the occurrence of which in fields is directly due to the sowing of their seeds mixed with grain or other crop seeds. Among the most injurious annual grasses are Wild Oats and the Poison Darnel [Plate 51]. Chess, *Bromus secalinus*, L., is a biennial and is often abundant in fields of fall wheat, particularly where there is a thin stand of the grain plants. Of perennial grasses which are troublesome, the worst are: the notorious Couch or Quack Grass [Plate 46] of the East; and the Sweet Grass [Plate 49], Skunk-tail Grass [Plate 47] and Spear Grass, *Stipa spartea*, Trin., of the West.

CHESS, *Bromus secalinus*, L., is an introduced biennial which is hardier than wheat, and where young plants of fall wheat have been killed out by the winter, Chess plants growing among the wheat from seeds sown with the grain, are seldom injured, but flourish to such an extent that some farmers have been led to the erroneous conclusion that Chess has originated from wheat plantlets which have been injured in various ways. It has, however, been proved conclusively that Chess is an entirely distinct grass which can grow only from its own kind of seeds; moreover, these seeds [Plate 56, fig. 78—natural size and enlarged 4 times] always have upon them a husk with a row of bristles down each side of the groove, by which as well as by their shape they may be easily distinguished from those of wheat. All doubters are recommended to dig up some plants of Chess as soon as they are recognizable in their fields, when they will find that the seeds from which the Chess plants began to grow the previous autumn, are still attached to the roots and that these are very different from grains of wheat.

Plate 46

COUCH, QUACK OR SCUTCH GRASS
(Agropyrum repens, L.)

PLATE 46.

COUCH OR QUACK GRASS, *Agropyrum repens* (L.) Beauv.

Other English names: Scutch, Twitch, Quitch.
Other Latin name: *Triticum repens*, L.

Introduced and native. Perennial by very wide-spreading but shallow fleshy rootstocks, forming large, matted beds. Flowering stems rather freely produced, smooth above, downy on the leaf sheaths below. Flowers in 3 to 7-flowered spikelets, forming a narrow spike with the spikelets lying flatly against the central stalk. Leaves grayish green, rather distinctly ribbed, and more or less hairy. Seed in the husk [Plate 56, fig. 79—natural size and enlarged 4 times] about ⅜ of an inch long, slender, 5 to 7-nerved, usually with an awn ⅛ inch long at the tip; the seed itself is shaped like a small grain of wheat $\frac{3}{16}$ of an inch long with wide open crease, the basal germ end pointed, and at the top a blunt fuzzy tip.

Time of Flowering: About the end of June; seeds ripe July.

Propagation: By seeds and extensively creeping rootstocks near the surface of the ground. When broken by plough or cultivator, every piece of the rootstock is capable of forming a new plant; such pieces may be carried from field to field on farm implements.

Occurrence: In all kinds of soil. The eastern form with bright green leaves, which is probably the European plant, that has been introduced, is abundant east of the Prairie Provinces, and also in a few localities in Manitoba and the North-west. The native form with very grayish green foliage, named *Agropyrum glaucum*, R. & S., var. *occidentale*, V. & S., is a far less aggressive weed, even where both forms are growing close together.

Injury: A most persistent weed in all deep-ploughed land, and in all crops, with great power of spreading and choking out other plants. The seeds are a very common impurity among seeds of the coarser grasses and in oats.

Remedy: Shallow ploughing in hot weather is essential in clearing land of this well known weed. Thorough harrowing after ploughing will drag out many of the fleshy rootstocks, which will soon dry up in the sun and can be burnt.

Rape sown after the land has been harrowed two or three times and well cultivated, is one of the best cleaning crops for late sowing; the seed should be sown 4 pounds to the acre in drills 26 inches apart, and the field kept clean with horse hoe, and afterwards with more or less hand hoeing if required. The land may be put under another hoed crop, corn, potatoes or other roots the following year.

It is recommended by some who have had experience in fighting Quack Grass, that in badly infested fields, the land be ploughed shallow late in the autumn and well cultivated to expose the rootstocks to the action of the frost. In the spring, re-plough shallow, and keep the ground stirred frequently enough to prevent new growth till midsummer, then sow a smothering crop, such as buckwheat or millet, which will choke out the weakened plants. It may be necessary sometimes to follow the above treatment with a hoed crop.

PLATE 47.

SKUNK-TAIL GRASS, *Hordeum jubatum*, L.

Other English names: Skunk Grass, Squirrel-tail Grass. Wild Barley, Tickle Grass, and, inaccurately, "Fox-tail."

Native. Perennial, not flowering the first year, forming tufts 8 to 12 inches high. Leaves grayish green. Flowers in beautiful silky bristly heads 3 to 4 inches long, pale yellowish green, often tinged with red. When ripe the spikes break up into 7-awned clusters of three flowers, the central one of which is long-awned and fertile; this produces a slender sharp-pointed seed; on each side of this and attached to it at the base are two abortive florets, each with three shorter awns than the central one; both the sharp seed and the awns are barbed upwards.

Time of Flowering: July; seeds ripe July to August.

Propagation: By seeds. This grass is frequently stated in works in which it is mentioned, to be an annual or a biennial; but all of the plants which I have grown for many years at Ottawa from western seed during the past twenty years, are certainly perennial, forming large tufts, but sending out no running rootstocks.

Occurrence: From Lake Superior westward, particularly in alkaline soil, where other better grasses cannot thrive.

Injury: This native grass is a very serious enemy of western stockmen, and is a source of much injury to horses, cattle and sheep. The barbed seeds and awns, when taken into the mouth, penetrate the soft tissues, causing annoying irritation and inflamed ulcers, which make the animals bite their tongues and lips so that these are frequently badly lacerated. They also work down beside the teeth, causing great inflammation and eventually swellings which have sometimes been taken for the disease known as Lumpy Jaw or Big Jaw (actinomycosis). The awns are also said by Prof. Aven Nelson (Wyoming Exp. Station, Bull, 19, 1894), to "work into the wool about the eyes of sheep, and then into the tissues surrounding the eye, and even into the ball itself, and in many instances causing total blindness." He also cites one case where this injury happened to the whole of a bunch of calves.

Remedy: Mr. T. N. Willing, of Regina, Sask., who has carefully studied the weeds of the North-west, sums up the best methods of dealing with Skunk-tail Grass as follows in his Bulletin No. 16, "Hints for the Grain Grower," 1905: "There is no difficulty in eradicating this grass from any land which can be ploughed, as the usual method of breaking in June will destroy it. It gives most trouble in waste places where it ripens its seed, which is spread abroad in every direction by wind and water. It grows freely about the edges of hay sloughs on the prairie, and is generally ripe before any hay is cut. The remedy in this case would be cutting before the seeds were formed. In a wet season, probably a second cutting would be necessary to prevent any seed ripening. If this course were continued for a few seasons, the pest would die a natural death, but it is the usual practice not only to cut too late, but also to avoid cutting the borders of sloughs in dry seasons when the grass is thin. Such methods favour the spread of this

Plate 47

SKUNK GRASS. WILD BARLEY OR SQUIRREL-TAIL GRASS

(*Hordeum jubatum*, L.)

objectionable grass. When fields of Awnless Brome Grass are badly infested, it is best to break and backset them, and then take a crop of grain before re-seeding, or the field may be burned over in the fall to destroy such seed of the Wild Barley as may have fallen, and early in the following spring plough the sod shallow, and then harrow and roll. In this way the Brome Grass may be renewed without re-seeding, and most of the weed will have been destroyed.

"Excessive irrigation is said to favour the growth and spread of this weed. Ditches and roads should be kept free of it."

In addition to the above good advice, it may be mentioned that when hay is found to be badly mixed with Skunk-tail Grass having ripe or hard seeds, owing to the light feathery nature of the dry heads, the greater part of them may be removed by tossing the hay lightly with a pitchfork on a windy day. The heads can easily be disposed of afterwards by raking together and putting a lighted match to the pile.

Skunk-tail Grass is a true Barley, and when young makes excellent feed both green and as hay.

SHORT-AWNED SKUNK-TAIL GRASS. Growing with the ordinary long-awned and pale-coloured form is an easily detected variety with much shorter awns to the seeds, taller and more erect stems, and the spikes which are slenderer and more drooping much more tinged with red. The habits of the two grasses are identical, and although generally growing together and equally common, one form as a rule preponderates in a locality.

PLATE 48.

WILD OATS, *Avena fatua*, L., var. *glabrata*, Petermann.

Other Latin names: *Avena fatua*, L., var. glabrescens. Cosson, and *Avena strigosa*, Schreb., of Canadian writers.

(Noxious: Dom., Ont., Man., N.W.)

Introduced. Annual, smooth, 2 to 4 feet high, growing in erect tufts. Plant closely resembling in general appearance some varieties of cultivated oats. Panicle loose and open, spreading in all directions, 6 to 12 inches long. The empty glumes or outer pair of scales, which enclose the 2 to 3-flowered spikelets, green, herbaceous and thin, as in the cultivated oats, both of about the same length. The flowering glumes or husks of the seeds [Plate 56, fig. 80—natural size and enlarged twice] hard and horny, bearing many short stiff bristles particularly about the base, rounded on the back, tapering to a slender papery divided tip, 7-nerved, the nerves roughened with minute teeth, and the central one running out from the middle of the husk into a stiff twisted bristle-like awn, nearly an inch long, which in the ripe seeds is bent at a right angle a little below the middle. The 2 or 3 florets or oats in a spikelet vary much in size and colour, the lowest one being much larger than the upper ones.

There are two varieties of the Wild Oat found as grain-field weeds in Canada. The type of the species, *Avena fatua*, L., which has rather larger, darker brown and much more bristly oats, is found in the eastern provinces; and the variety *glabrata*, with smaller, smoother, gray or olive-brown oats, with comparatively heavier kernels, is the prevailing form in the West. The identification of the variety *glabrata* was kindly made for me by Prof. C. V. Piper, of Washington, D.C. The White Wild Oats sometimes found are merely albinoes of these two varieties and do not always come true from seed. Carefully selected white oats under cultivation produced many dark-seeded plants and dark seeds gave several white oats. Both the eastern and western Wild Oats may nearly always be distinguished from cultivated oats by their earliness and the marked irregularity of ripening their seeds, the top oats frequently being ripe and shelling out long before the florets at the bottom of the panicle are ripe. Wild Oats contain smaller kernels, have always awns and some bristles at the base. The slanting horse-shoe shaped scar at the base of the seed is densely bristly. although these bristles are easily broken off, as also, with less frequency, is the scar itself, sometimes making the certain identification of some seeds difficult when found among threshed grain. The plate given herewith of the variety *glabrata* was drawn by Mr. Criddle from a plant found in Manitoba. When the panicles first appear from the sheaths, they are much more contracted in form.

Wild Oats may be found in flower by the end of June, and some seeds are ripe by the middle of July.

Propagation: By seeds only. Plants cut off when in flower throw up secondary flowering stems very quickly.

Occurrence: In all parts of the country, the seeds being widely distributed with all kinds of cereal grains and also carried from farm to farm in imperfectly cleaned threshing machines.

Plate 48

WILD OATS

(*Avena fatua* L. var. *glabrata* Petermann.)

Injury: Wild Oat plants are hardier and able to withstand injury much better than the cultivated varieties of oats. The seeds, unlike those of the latter, can remain in the soil in a vital condition not only over our severest winters, but for several years. On account of their irregular and early ripening, many seeds ripen and shell out before the grain crops in which they are growing are ready to cut. Thus the land becomes infested with the seeds, which continue to appear in succeeding crops for many years.

Remedy: Sow clean seed grain. Wild Oats, if in the land, must be grown out of it by adopting some method by which the seeds are made to germinate and the young plants are destroyed before they have ripened their seeds. In the East, a short rotation, with seeding down for hay and pasture at regular intervals, should be adopted. In the West, where grain is grown almost to the exclusion of all other crops, the land should be cultivated directly after harvest to cover up seeds on the surface. Many of these will germinate in autumn and will be killed by the winter cold. Early the next spring more seeds will germinate; these must be cultivated down as soon as there is a good growth; plough about 1st June and sow early oats or barley to be cut for green feed as soon as the heads appear. The land may then be ploughed and cultivated twice or as often as necessary to kill the weeds before winter, or a second cutting of green feed may be taken off before ploughing down the stubble. Instead of growing a green feed crop, some farmers sow early barley and leave for seed; but care must be taken that no Wild Oats ripen. If plenty of cattle are available, the Wild Oats may be kept fed off as they come up on stubble land, and seeding down will help very much to keep them in check.

PLATE 49.

SWEET GRASS, *Hierochloa borealis,* R. & S.

Other English names: Indian Hay, Vanilla Grass, Seneca Grass, Holy Grass.

Other Latin names: *Holcus odoratus,* L.; *Hierochloe borealis,* Roem. and Schultes; *Savastana odorata* (L.) Scribn.

Native. Perennial, deep-rooted, with wide-spreading white rootstocks which produce in summer many barren shoots with long flat shining leaves over a foot in length, of a deep green. Flowering stems thrown up very early in spring, the first flowers opening when the stems are only a few inches out of the ground. Panicle pyramidal, 1 to 2 inches high, loose during flowering, with spreading branches, contracting as the seeds ripen, when the stems are 12 to 18 inches high; the sheath with its short blade below the middle. Spikelets drooping, with shining papery outer glumes which are yellowish, tinged with purple, 1-seeded but 3-flowered, two male flowers with 3 stamens between downy ciliate-margined scales, and one fertile flower inside 2 smooth scales with 2 anthers and a double plumose pistil. When the seeds are ripe, the whole panicle becomes dark golden brown. Seeds [Plate 54, fig. 39—natural size and enlarged 8 times] enclosed in the inner scales, small, $\frac{1}{10}$ of an inch, oblong. Whole plant sweetly aromatic, with the fragrant principle of the Tonka bean and Sweet Clover (Coumarin).

Time of Flowering: April to May. Fruit ripe by the beginning of June.

Propagation: By seeds and running rootstocks.

Occurrence: Rare in the eastern provinces and growing in damp places by streams and rivers. In the West, in all kinds of soil, extremely abundant and very difficult to eradicate.

Injury: This early ripening and deep-rooted grass is very persistent in the rich lands of the West, where it smothers out all kinds of crops.

Remedy: Mow and burn before summer-fallowing so as to avoid ploughing down ripe seeds. Ploughing for summer-fallows must be deep. Good results have been secured in Manitoba by ploughing in spring when the Sweet Grass is in flower and then seeding down heavily at once.

This grass is often incorrectly spoken of in the West as "Quack" or "Couch Grass," quite a different grass, which flowers at the end of June, roots near the surface of the soil, and can be killed by shallow ploughing followed by frequent cultivation. Sweet Grass flowers in April, and shallow ploughing merely stimulates its growth.

Plate 49

SWEET GRASS
(Hierochloa borealis, R&S.)

Plate 50

GREEN FOXTAIL
(*Setaria viridis*. *Beauv.*)

PLATE 50.

GREEN FOXTAIL, *Setaria viridis*, Beauv.

Other English name: Pigeon Grass.

Other Latin names: *Chætochloa viridis* (L) Scribn.; *Ixophorus viridis* (L) Nash; *Chamæraphis viridis*, Porter.

Introduced. Annual. Stems several, erect, simple or branched from below, 1 to 2 feet high, leafy. Panicle condensed into a cylindrical compound spike. Spikelets single-flowered, or with a perfect flower and a neutral flower beside it, inside 3 empty glumes or scales, awnless, but with a cluster of 3 to 6 persistent green bristles below the florets on the short peduncle. The tough hard husk of the seed is about $\frac{1}{12}$ of an inch long, oval, with the glume or outer scale rounded and folded over the highly polished rounded edges of the pale or inner scale, which is flattened in the middle. Both scales of the husk are roughened crosswise with narrow ridges. The colour is very variable, according to the degree of ripeness—yellow, gray, brown or purplish, the darker seeds mottled with darker spots. The seed itself is greenish white, convex on the outer germ-bearing face and flattened on the inner face. [Plate 54, fig. 40—natural size and enlarged 8 times.]

Time of Flowering: June to September; seeds ripe by July.

Propagation: By seeds.

Occurrence: Abundant in Eastern Canada; as yet only occasional in the West.

Injury: The seeds are one of the commonest impurities in clover and grass seeds.

Remedy: A common and abundant weed in all crops on land not worked under a short rotation. Easily killed when young by smothering with earth in the cultivation of hoed crops. Late plants in hoed crops, after cultivation stops, should be pulled by hand.

YELLOW FOXTAIL, *Setaria glauca*, Beauv. Another grass of the same family and very similar to the above is the Yellow Foxtail. The branches, however, are more decumbent and spreading, the whole plant rather larger and more succulent, the spikes less compound and slenderer, with larger seeds $\frac{1}{8}$ inch, and the bristles are distinctly yellow. The seeds are almost equally common with those of the Green Foxtail in commercial seeds.

PLATE 51.

COMMON DARNEL, *Lolium temulentum,* L.

Other English names: Poison Darnel, White Darnel, Ivray, Poison Rye-grass, Bearded Darnel.

Other Latin names: *Lolium arvense,* With.; *Lolium temulentum,* L., var. *a genuinum,* Sm.; *Lolium temulentum,* L., var. *b arvense,* Sm.

Introduced. Annual, smooth. Stems 2 to 4 feet high, simple, leafy. Flowers in a spike 6 to 10 inches long; somewhat resembling that of Couch Grass, but having the edges of the spikelets, instead of the broadsides as in Couch, resting against the stalk; spikelets 3 to 7-flowered, solitary, sessile and alternate, with their edges fitting tightly into grooves on either side of the stalk; each spikelet in the axil of a long rigid strongly nerved, persistent glume or empty scale, which merely equals or is much longer than the spikelets. Leaves 6 to 10 inches long by ¼ inch wide, rough above. Seeds swollen, nearly straight on the outer face, much swollen on the inner, with a deep wide groove, the inner scale of the husk with a wing-like keel on each side, minutely bristly on the edges, but not coarsely bristly ciliate as in Chess, *Bromus secalinus;* outer scale hard and flinty as in the chaff of wheat, and either with a long awn in the variety *genuinum,* which is the more widely distributed form of this weed in Canada, or entirely without awn in the variety *arvense,* which has recently appeared in great abundance at certain points in the Red River Valley in Manitoba, where the seed was spoken of as "False Barley" when first detected. The true seed after the husks have been removed is greenish brown often tinged with deep purple.

Time of Flowering: July; seeds ripe August.

Propagation: By seed.

Occurrence: Occasional, generally in wet land; abundant recently in parts of the Red River valley, Manitoba.

Injury: The husks cover the seed very tightly, the inner scale being adherent to it; in that condition the seeds are very much of the same size as small grains of wheat; they are therefore very difficult to separate from that grain.

The seeds of Darnel are widely reputed to be poisonous; but there seems to be some doubt upon the subject. In "The True Grasses," by Eduard Hackel, as translated by Prof. F. L. Scribner and Effie A. Southworth, is the following: "A weed among grain crops; troublesome in wet years. The grain contains a narcotic principle (Loliin) soluble in ether, which causes eruptions, trembling and confusion of sight in man and flesh-eating animals, and very strongly in rabbits; but it does not affect swine, horned cattle or ducks."

Prof. E. M. Freeman, of the University of Minnesota, has made some important investigations into the question of the origin of darnel poisoning, and in commenting on the discovery by P. Guerin, of Paris, France, of a fungus in the seeds of Darnel, to which he attributed the poisonous effects, says as follows under date April 3rd, 1906: "In California Darnel is exceedingly abundant, and is known as Chess or Cheat. It sometimes

Plate 51

COMMON DARNEL
(*Lolium temulentum L.*)

in wet situations or poorly drained regions, constitutes a very considerable portion of the grain." "From the literature, which I have examined somewhat extensively, I learn that the concensus of opinion in Europe is that the plant is poisonous. I have shown, however, [Phil. Trans. Roy. Soc. Lond. Sec. B. Vol. 196, pp. 1-27, 1903.] that there are two races of the plant, one with a fungus, and the other without a fungus, and there is apparently no transference of the fungus from one race to the other. If the seeds are really poisonous, it may be that those with the fungus are poisonous, while those without fungus are not. I have attempted recently to determine this, but have failed to get any conclusive results." "It is well known that some Europeans doubt the toxic action of these grains, and it is worthy of note that *Lolium temulentum* very commonly carries ergots. Kobert and other European investigators ascribe the toxic effects chiefly to the presence of ergot. I have personally attempted also to obtain evidence of toxicity by feeding the seeds with the fungus and without the fungus to rabbits, mice and guinea pigs. with absolutely no effect. As far then as my own experience is concerned, I have been unable to prove the toxicity of the Darnel; but I must admit that these experiments have not been at all extensive and are undoubtedly insufficient to serve as a basis for far-reaching conclusions. I have recently obtained information from California, which tends to show, however, that the plant in that region is not at all feared. In fact, I have letters from men interested in the milling of wheat in California who inform me that the grain is sometimes intentionally mixed with barley for feed, and I can get absolutely no record of any toxic action of the plant from California. As I have stated above, Darnel is exceedingly common in that State."

The above gives, I believe, a summary of this subject up to the present time.

Remedy: Sow clean seed. Should Darnel increase very much, doubtless some improvement will be made in fanning mills by means of which the seed will be separated from the grain, as has recently been done in the case of Wild Oats.

PLATE 52.

ERGOT ON COUCH, RYE AND TIMOTHY,

Claviceps purpurea (Fr.) Tul., and other species.

(Noxious: Dom.)

There are often found in certain seasons among the grains of rye, rarely among those of wheat, and abundantly among the seeds of some grasses, dark-coloured solid bodies of characteristic form in each species of grain or seed. These are of a doughy consistency, and when broken are purplish white inside. They are the storage organs or resting stage of a parasitic fungus or perhaps the several species of fungi belonging to the genus *Claviceps*. Each of these solid bodies is called a *sclerotium* (plural *sclerotia*) derived from a Greek word *skleeros*, hard or dry, in allusion to the nature of these bodies. They are practically masses of the vegetative system or "spawn" of the fungus in a resting condition, but capable of growth in spring when placed under favourable conditions of warmth and moisture, such as they get when sown with crop seed or when lying on the ground at the bases of the stems on which they were formed the previous summer. At the proper time in spring very small toadstool-like bodies on violet stalks with round orange-coloured heads about the size of mustard seeds, are produced from the sclerotia lying on the ground, which develop enormous numbers of microscopically small spores (organs analogous to the seeds of higher plants). These are produced at the time that grasses and grains are in flower. The minute spores, carried by currents of air or by insects, lodge in the flowers of the grasses and begin to grow; in a short time they completely destroy the seed and form from them the sclerotia. These vary in shape according to the plant they attack. In rye and Couch Grass they are long and horn-like, over 1 inch in length, frequently much larger than shown in our figure of Couch Grass. In wheat and Wild Rice they are shorter and thicker; in timothy and June Grass very small, ⅛ of an inch long, slender and almost black. During the summer, summer spores are formed on these horns, and also at the same time a sugary secretion very attractive to insects which, coming to the infested plants, carry off on their bodies many of the summer spores to the flowering heads of other grasses they visit, and in that way spread the infection. Late in summer the production of summer spores stops, and the sclerotia or storage organs begin to store up a special kind of starch found only in fungi and known as fungus starch, as well as oils to serve as food for the growth of the fruiting organs to be sent out the following spring; they then harden up, turn dark purple in colour and fall to the ground or are carried with the grain or hay.

The sclerotia are common on a great many grasses and particularly on rye, wheat, barley and Wild Rice, as well as on Western Couch Grass and other prairie grasses cut for hay. They all contain an alkaloid and other violent poisons. Some are used in medicine under the name of "Ergot of Rye." Bread made from flour containing ergot may cause a serious disease known as ergotism in those that eat it; and animals which feed on grain or hay containing ergot, may also be severely poisoned, as is sometimes the case on our western plains. One well-known result of cattle eating ergotised hay is abortion.

Plate 52

COUCH OR QUACK GRASS, RYE AND TIMOTHY
Attacked by Ergot

Hay containing much ergot should not be fed. In years of scarcity when such hay must be used, it should be well threshed before feeding, to dislodge the poisonous sclerotia. Grain containing ergot should be thoroughly screened before use and the sclerotia destroyed. Seed grain from an ergotised crop should not be used if any other grain can be procured, and all grain should be treated with formalin or bluestone before sowing as a precaution against ergot and smut.

Many writers have treated of Ergot, as Berkeley, Cooke, Worthington Smith, Kerner & Oliver, Bessey, Halsted and many others. One of the best and most recent articles of a popular but scientific nature is by Prof. E. M. Freeman in his excellent "Minnesota Plant Diseases" (1905). Ergot and Ergotism are also treated of in all the leading encyclopædias.

Plate 53

1
Wormseed Mustard
2
Green Tansy Mustard
3
Tumbling Mustard
4
Wild Mustard
5
Shepherd's-purse
6
Stink Weed
7
Peppergrass
8
Spurrey
9
Bladder Campion
10
Sticky Cockle
11
White Cockle
12
Chickweed
13
Grass-leaved Stitchwort
14
Mouse-Ear Chickweed
15
Purslane
16
White Evening-primrose
17
Common Evening-primrose
18
False Tansy
19
Mayweed
20
Ragwort

Plate 54

21
Chicory
22
Field Sowthistle
23
Common Sowthistle
24
Prickly Sowthistle
25
Common Plantain
26
Pale Plantain
27
Hoary Plantain
28
Ribgrass
29
Bracted Plantain
30
Toadflax
31
Vervain
32
Dragonhead
33
Catnip
34
Clover Dodder
35
Lamb's-quarters
36
Red-root Pigweed
37
Lady's-thumb
38
Sheep Sorrel
39
Sweet Grass
40
Green Foxtail

Plate 55

41
White Anemone
42
Tall Buttercup
43
Golden Fumitory
44
Hare's-ear Mustard
45
False-flax
46
Ball Mustard
47
Spider-flower
48
Cow Cockle
49
Purple Cockle
50
Wild Tares
51
Prairie Rose
52
Cow-bane
53
Gum-weed
54
Poverty-weed
55
False Ragweed
56
Black-eyed Susan
57
Tall Ragweed
58
Common Ragweed
59
Perennial Ragweed
60
Wild Sunflower

Plate 56

61
Yarrow
62
Burdock
63
Knapweed
64
Canada Thistle
65
Fall Dandelion
66
Orange Hawkweed
67
Prickly Lettuce
68
Blue Lettuce
69
Blue-weed
70
Field Gromwell
71
Blue-bur
72
Field Bindweed
73
Hairy Mint
74
Russian Pigweed
75
Wild Buckwheat
76
Curled Dock
77
Sun Spurge
78
Chess
79
Quack
80
Wild Oats

INDEX.